2 Receptors

Edited by
P. Cuatrecasas
Wellcome Research Laboratory,
Research Triangle Park, North Carolina

and
M. F. Greaves
ICRF Tumour Immunology Unit,
University College London

and Recognition
Series A

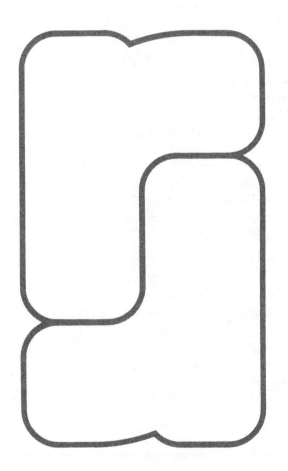

LONDON

CHAPMAN AND HALL

A Halsted Press Book

John Wiley & Sons, Inc., New York

First published 1976
by Chapman and Hall Ltd
11 New Fetter Lane, London EC4P 4EE

© 1976 Chapman and Hall

Typeset by Josée Utteridge of Red Lion Setters
and printed in Great Britain at the University Printing House, Cambridge

ISBN 0 412 13810 7 (cased edition)
ISBN 0 412 14280 5 (paperback edition)

Distributed in the U.S.A. by Halsted Press,
a Division of John Wiley & Sons, Inc., New York

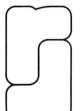

Contents

Preface

Volume 2 in the 'Receptors and Recognition' series contains a mixture of review articles which may at first sight appear quite unrelated. What connection can there be for example between incompatibility in flowering plants and catecholamine receptors? Why should a pharmacologist or a botanist read more than one of these articles? It is, however, the diversity of cellular and molecular systems described in this volume that illustrates the philosophy of the series as a whole. The individual chapters describe in clear terms, for the relative non-specialist, the basic elements of recognition events in distinct biological systems of cellular interaction and regulation. We are using this series to promote the view that it will be both instructive and entertaining to compare recognition events in widely different biological systems and to search for evolutionary links, common mechanisms or ground rules. This approach should ideally lead to a cross-fertilisation of concepts, ideas and methodologies.

We believe we have succeeded in persuading many first class biologists in this and companion volumes to elucidate the central features of the systems on which they are authorities. The articles are written in an imaginative way and we hope that the audience has appropriate, complementary receptivity. We hope, for example, that biologists interested in cell regulation will appreciate from Gomperts' article the likely central regulatory role of calcium ions. We urge those interested in both vertebrate reproduction and in mammalian histocompatibility system to read Lewis' fascinating article on the recognition systems developed by plants to ensure heterozygosity. Although the molecular mechanisms involved in plant incompatibility reactions are unknown as yet, the similarities of their genetic control and overall structural design to more familiar animal cell interactions are quite striking. This topic will be elaborated further by Dr Marjorie Crandall in Volume 3 of 'Receptors and Recognition' ('Mating type interactions in microorganisms.').

Givol and Levitski have contributed excellent reviews on molecular aspects of 'recognition' molecules — antibodies and catecholamine receptors respectively. Antibodies are in a sense the clinical example of complementary recognition molecules but, as Givol lucidly describes, it is only quite recently that a detailed picture of the active or combining sites of antibodies has begun to emerge along with some understanding of the genetic control of immunoglobulin structure. Catecholamine receptors have proved extremely difficult to 'pin down' and distinguish from other less specific binding sites on cell surfaces. More careful and sophisticated methods are now available to study these receptors and Levitski's article describes a marvellous piece of detective work on this hitherto elusive receptor.

Finally, Maria De Sousa's review on cell traffic reminds us that whole cells as well as their membranes are not all stationary. Cell movement, seeding, migration, and in a few cases recirculation, play a vital role in the development and function of many cells. Much of this pattern of cellular behaviour appears at least superficially to be rather frantic and governed by only physical constraints. However, there is little doubt that although cells may not know where they are going they often do seem to know where they are. Cell surface structures and cell interactions undoubtedly play a key role in governing the nomadic behaviour of cells. The state of the art in this field is still largely at the descriptive or phenomenological stage but the development of good model systems should soon lead to a clear understanding of molecular mechanisms. At the same time it is possible that this type of dynamic cell behaviour will be much more difficult to unravel in molecular terms than those cell surface recognition events involving diffusible regulatory ligands such as the catecholamines. The latter are themselves available in purified forms and amenable to radio-labelling which greatly facilitates receptor identification.

October 1976   P. Cuatrecasas
                                                        M. F. Greaves

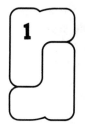

1

A Structural Basis for Molecular Recognition
The Antibody Case

DAVID GIVOL
Department of Chemical Immunology,
The Weizmann Institute of Science,
Rehovot, Israel

Abbreviations

Ig — immunoglobulin; H — heavy chain; L — light chain; V — variable; C — constant; L1, L2, L3 and H1, H2, H3 — the hypervariable segments of L and H chains respectively; BADE — N-bromoacetyl-N'-Dnp-ethylenediamine; BADB — α-N-bromoacetyl-γ-N-Dnp-α,γ-diamino-L-butyric acid; BADO — α-N-bromoacetyl-δ-N-Dnp-L-ornithine; BADL — α-N-bromoacetyl-ϵ-N-Dnp-L-lysine; NAP-Lys — 4-azido-2-nitrophenyl-L-lysine; Dnp-N$_3$ — 1-azido-2,4-dinitrobenzene; Dnp-AD — dinitrophenyl-alanyl-diazoketone; DPPC — p-diazoniumphenyl-phosphorylcholine; MNBD — m-nitrobenzenediazonium; BAAT — bromoacetyl-arsanyl-azotyrosine; NAP — 4-azido-2-nitrophenyl.

Acknowledgement

Supported by Grant 4 R01 AI 11453 from the National Institutes of Health. I thank Dr I. Schechter for helpful discussions concerning combining sites.

Receptors and Recognition, Series A, Volume 2
Edited by P. Cuatrecasas and M.F. Greaves
Published in 1976 by Chapman and Hall, 11 New Fetter Lane, London EC4P 4EE

INTRODUCTION

Molecular recognition stands at the roots of biology. All living processes are dependent on enzymatic reactions, which in one form or another are conducted in an orderly way. The first step in all these processes is the preferential binding of one molecule by another, which is an expression of molecular recognition and, the spatial folding of proteins is the basis for their ability to recognize specific structures via combining sites. These binding sites are designed to provide the necessary geometrical complementarity to fit their specific ligands, the strength of the binding being due to various interactions between the favorably fitted ligand and various groups in the protein. Recognition via combining sites is a property of enzymes, antibodies, hormones, receptors, lectins, and perhaps also other classes of proteins, but it is best expressed in its diversity and specificity in enzymes and antibodies. These two classes of proteins can discriminate between numerous compounds. It is possible to isolate or to select enzymes of almost any structure, and it is similarly possible to induce the formation of antibodies to any small or large molecule. It is, however, impossible to find a structural common denominator for all enzyme combining sites. In contrast, antibodies form a unique system where the multiple of different combining sites can by analyzed on a background of similar amino acid sequence as well as a similar three-dimensional structure. The rapid development of chemical and crystallographical analysis of antibodies now permits an understanding of the structural basis of antibody specificity, and furthermore it provides a beautiful example of the possibility of generating many specificities within one general framework.

Another unique aspect of the immune system is its inducibility; i.e. the formation of antibodies is induced by antigens. The recognition of the antigen is by cell membrane associated antibody like receptors. The cellular receptor for the antigen is, at least on B lymphocyte, identical to the antibody later produced by the cell. This identity between cellular receptor on the cell membrane, on the one hand, and the cell product, on the other hand, adds to the interest in

3

understanding the structure and function of antibodies since it should provide some clues for ways of analyzing the behavior of cell receptors and cell triggering. This contrasts with other inducible systems in which there is no relationship between the receptor for the inducer and the product made by the cells as a result of the inductive process.

1.1 MOLECULAR RECOGNITION BY ANTIBODIES AND ENZYMES

1.1.1 Comparison of the two systems

If we consider antibodies and enzymes as a collection of recognition molecules, we will observe several noticeable differences between them. Enzymes are designed to bind ligands and catalyze a chemical reaction, the binding and catalytic sites being parts of an active site. Antibodies have the property of reversible binding of ligands; they are not capable of the formation or breaking of covalent bonds. Enzymes are produced in all cells; antibodies are the unique products of specific cells in the immune system, i.e. lymphocytes and plasma cells. It is interesting that lymphocytes can make more different combining sites, as antibodies, than all the other cells of the body *in toto*, since antibodies can be made against any particular foreign protein, including antibodies themselves. The formation of antibodies is induced by antigens, which are predominantly macromolecules that bind to lymphocyte receptors. Although formation of enzymes can also be induced, the inducers in this case are primarily small molecules. The structure of enzymes is widely variable for different enzymes. There is no similarity between an enzyme that acts on carbohydrates (e.g. lysozyme) and an enzyme that acts on polypeptides (e.g. trypsin). On the other hand, antibodies to carbohydrates and polypeptides cannot be resolved by any physico-chemical criteria, their general structure is very similar and they differ only in their combining site. The number of combining sites in an enzyme molecule may differ in different enzymes but generally one combining site is formed by one peptide chain. In antibodies there are always two combining sites per molecule (ten in IgM which is the pentamer of the basic subunit) and each combining site is formed by two dissimilar peptide chains. Random pairing between these two chains could give rise to a large number (n^2) of different combining sites with a much smaller number (n) of the two chains. This reduced the gene load necessary to code for such a large variety of proteins. In antibodies the peptide chains can be divided into a constant (C) region whose amino acid sequence is identical for all antibodies of one class, and a variable (V) region whose sequences differ from one antibody to another. In the lymphocytes these two portions of the peptide chain are being synthesized as one chain from one growing point by a single mRNA molecule. There is, however, solid genetic evidence that this peptide chain is coded by two separate genes. In enzymes and all other proteins the rule is one gene to one polypeptide chain. The two genes to one peptide chain mechanism in antibodies is probably being used also for the switch in synthesis from one immunoglobulin class (like IgM) to another (like IgG) without changing antibody specificity. In spite of the above differences it is interesting to compare the architecture of the combining sites of antibodies and enzymes.

1.1.2 General features of combining sites

Biological recognition in most cases can be analyzed by studying the interaction of macromolecules with small molecules. The strength of this interaction is measured in terms of affinity or the association constant K, and this is related to $\Delta F°$, the standard free energy of binding, by the equation $-\Delta F° = RT\ln K$. It is useful to remember that a 10-fold increase in the association constant K is reflected (at $27°$) in an approximately 1.4 kcal decrease in $\Delta F°$. Important progress in understanding the interaction between enzymes and substrate was made by the introduction of the concept of subsites by Schechter and Berger [1]. By using a series of oligopeptides with different sequence or optical configuration as either substrates or inhibitors of papain, it was shown that the combining site of the enzyme is larger than was previously thought, and can be divided into subsites. A subsite was defined as the region on the enzyme which interacts with one amino acid of the substrate [2, 3]. In the enzyme–substrate complex, the substrate was lined up in the combining site such that the group being hydrolyzed always occupied the same place, i.e. the catalytic site. Subsites will be denoted $S_1 - S_n$ or $S'_1 - S'_n$ if they are accommodating amino acids of the substrate which are respectively towards the N- or C-terminal side of the bond cleaved in the substrate. For example, papain was found to have seven subsites, four on the N-terminal side and three on the C-terminal side of the catalytic site (Fig. 1.1). A similar type of analysis was performed also on carboxypeptidase [4] and elastase [5], and was extended to enzymes that split

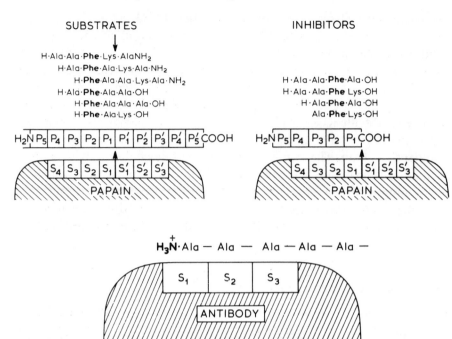

Fig. 1.1 Schematic representation of subsites in enzyme–substrate and enzyme–inhibitor complexes in the case of papain [1], and its analogy with antibody–hapten complex [14]. Reproduced with permission.

polysaccharides such as hen egg-white lysozyme [6] and different amylases [7], and to enzymes that split nucleic acids such as staphylococcal nuclease [8]. In the case of lysozyme a particularly useful comparison could have been made between the study of substrate binding in solution and the X-ray analysis of enzyme–substrate crystals which clearly delineate the substrate in these subsites [9, 10]. Thus, the catalytic site of lysozyme, consisting of the carboxylate of Asp 52 and the carboxyl of Glu 35, is situated between S_1 and S_1' which are subsites D and E respectively in the three-dimensional model of lysozyme [9].

An important feature of the subsites concept is that, to a first approximation, the binding forces operating in individual subsites are additive, such that partial $-\Delta F°$ can be attributed for each subsite and their sum will be the $-\Delta F°$ of the substrate binding. This permits mapping of the combining site as 'viewed' by the substrate. For a pair of petides or saccharides (substrates or inhibitors) in which only one residue is replaced, the difference in standard free energy $\Delta\Delta F (\Delta^2 F)$ can be calculated from the difference in their inhibition constants [2], binding constants [6], or rates of hydrolysis [2, 7]. Hence a $\Delta^2 F$ can be assigned to a subsite.

An interesting aspect of this mapping was that there is no linear order in the contribution to $-\Delta F^\circ$ by the various subsites. For example, in the case of papain a change of Gly to Ala in S_1 (i.e. an additional methyl group) adds about 1 kcal to the binding energy. Replacing L-alanine by L-leucine at S_1 will add to $-\Delta F^\circ$ about 1 kcal/mole, and so does L-phenylalnine. However, binding in S_2 is much stronger than in S_1 although it is further removed from the catalytic site. A phenyl side-chain in S_2 will contribute 3 kcal/mole whereas in S_1 it will contribute only 1 kcal as compared with the methyl side-chain of alanine. In fact a hydrophobic residue like phenylalanine in a peptide substrate will always occupy S_2 and will thereby direct the enzymic cleavage to be next but one to this phenylalanine, towards the C-terminal side [3]. Similar studies with elastase have demonstrated eight subsites, three on the C-terminal side and five on the N-terminal side of the cleavage point [5]. It is interesting that subsite S_4, though far from the cleavage point, was found to contribute significantly more than other subsites to the binding energy as well as to the rate of hydrolysis. Thus Ala_4 - Lys-Phe has an affinity constant 110-fold greater than that of Ala_3 -Lys-Phe [5]. The contribution of binding energy by saccharides at different subsites of lysozyme was analyzed by Chipman and Sharon using binding data for 26 saccharides of different lengths [6]. Interestingly, in this case S_1 (subsite D) is not merely a non-interacting site but must lead to unfavorable interaction since the tetrasaccharide GlcNAc-MurNAc-GlcNAc-MurNAc is bound much more weakly than the trisaccharide GlcNAc-MurNAc-GlcNAc. This is probably due to the distortion of the saccharide in this subsite, which is related to the catalytic mechanism. The contribution of binding energy by other subsites is quite random, the highest energy being contributed by S_2 (subsite C). This subsite cannot accommodate MurNAc which was shown to be favorably bound by S_3 (subsite B or S_1 [6]. Comparison of three different amylases [7] again points to the fact that the contribution to binding energy by different subsites is quite random (Fig. 1.2).

The study of subsites in antibodies was first approached by Kabat by comparing the capacity of saccharides of increasing size to inhibit the reaction between antibodies and polysaccharides [11, 12]. In the dextran—anti-dextran system it was shown that the inhibition power increases with the increase in chain length, reaching an upper limit at the

size of isomaltohexaose. The $\Delta\Delta F°$ attributed to the increase in length
by each saccharide can be calculated; it was shown that the terminal
glucose contributes most to the $\Delta F°$, and each succeeding glucose contributes a smaller increment.

The mapping of antibody combining sites by division into subsites
was also studied by Schechter using antibodies against oligopeptides and
particularly anti-tetra-alanyl antibody [13, 14]. By systematic variations
in the structure of tetra-L-alanine where L-alanine was replaced with
D-alanine, glycine, or L-phenylalanine, the binding energy $(-\Delta^2 F)$ of the
methyl side-chain of alanine in each subsite could be estimated. It was
indeed found that the methyl group of Ala was bound with different
energies in different subsites. The site of anti-alanyl antibody was found
to be composed of three subsites, and the binding energy in individual
subsites was found to be additive. However, in contrast to the active sites
of enzymes, in the antibody combining site there is a vectorial decrease
of the binding energy and it always drops gradually from S_1 to S_n, S_1
being the terminal portion of the determinant, i.e. the residue most
distal from the antigen carrier. A schematic representation of some
enzyme and antibody sites bearing on this aspect of the subsite mapping
is given in Fig. 1.2. These differences in the distribution of bindings
energy in the active site of enzymes and antibodies were pointed out to
me by Dr I Schechter (unpublished data).

T That the terminal moiety of a haptenic determinant is the immuno-
dominant portion was first suggested by Landsteiner [15], and it was
found to hold true in most systems of sequential determinants that were
analyzed, in particular in the ABH system of blood groups and in all
haptens snalyzed [12]. An interesting example to emphasize the immuno-
dominance of the most distal part of the hapten was provided by
Schechter *et al.,* [16] who prepared antibodies to tetra-alanine which
was attached to the protein antigen in one case via its carboxy terminus
and in the other via its amino terminus. Both antigens induced good
antibody response but there was no cross reaction between anti-'amino-
alanyl' and anti-'alanyl-carboxyl'. The tetra-alanine peptide (Ala 4) will
bind to both antibodies. However, if the carboxy end is modified to an
amine (Ala_4 amide) it will bind only to anti-'amino-alanyl'. On the other
hand, if the amino terminal end is modified (N-acetyl Ala_4) it will only
bind to anti-'carboxy-alanyl'.

It should be emphasized that the examples cited above are convenient
for the analysis of the antibody site but they do not by any means cover
the entire range of antigenic determinants. Linear polysaccharides such
as type III and VIII pneumococcal polysaccharides contain non-terminal,

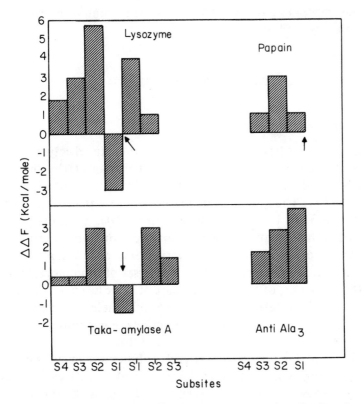

Fig. 1.2 Histograms of subsite affinities in enzymes and antibodies. Arrows indicate the point of cleavage in enzymes. Data for lysozyme are from binding of oligosaccharides of different lengths [6]. Data for taka-amylase are from kinetic analysis [7]. Data for papain are from inhibition constants of various peptide inhibitors of the hydrolysis of benzol-L-arginine ethyl ester by papain. The value represent the $-\Delta^2 F$ of Ac-Phe-Phe as compared with Ala-Ala-Ala [2]. Data for anti-poly-D-alanine antibody are from the use of various peptides to inhibit the precipitin reaction or the binding of the peptide D-Ala$_3$-Gly. Values represent the $-\Delta^2 F$ contributed by the interaction of each subsite and D-alanyl residue [14].

internal, antigenic determinants. Recently Cisar *et al.,* [18] re-analyzed the problem of terminal and non-terminal antigenic determinants in the dextran—anti-dextran system and found antibodies or myeloma proteins which are directed against non-terminal determinants. Also most of the native antigenic determinants of proteins are dependent on their native conformation, and it will be difficult experimentally to analyze them in terms of terminal and non-terminal moieties. However, it seems true that

the antibody recognizes the outer surface of the immunogen that was
used to elicit this antibody.

1.2 THE ANTIBODY MOLECULE

1.2.1 Polypeptide chain structure

All immunoglobulins have the same basic multi-chain structure. This is
a symmetrical molecule composed of two heavy chains (H) of molecular
weight 50 000 and two light chains (L) of molecular weight around
25 000 each [19]. The most unique feature of these polypeptide chains
is that only the first 110—120 residues from the NH_2-terminus (1/2 of
L and 1/4 of H) have sequences which vary from one protein to another,
whereas the rest of the chain has an invariant sequence at least within
the same immuglobulin class. The various classes of immunoglobulins
differ in the sequence, of the constant part of their heavy chain. The
classes IgG, IgA, IgM, IgD, and IgE differ also in their abundance in the
serum, in their non-specific biological functions (e.g. complement
fixation), and in their role as cell surface receptors.

The second notable feature is that each polypeptide chain is composed
of homologous segments (domains) of approximately 100 residues [20] :
(a) the homology segments show sequence homology with other constant
segments of both H and L chains, presumably as a result of gene duplication; (b) the homology segments contain one intrachain disulfide bond
between two half-cysteine residues which are about 65 amino acids apart;
(c) each homology segment folds, independently of the other segments,
as a globular domain; (d) homologous globular domains from different
chains interact by non-covalent bonds to yield a unit possessing a unique
biological function which may be retained even when such a fragment is
split and isolated from the entire molecule (Fig. 1.3). Indeed, the antibody
molecule can be split by various enzymes to yield the fragments Fab,
$(Fab')_2$, Fc, Facb, and Fv, as depicted in Fig. 1.3. Binding activity is
retained in all fragments which contain the N-terminal variable domain of
both H and L chains, whereas the binding of complement is retained only
in Facb indicating that it is located in $C_H 2$ domains [21].

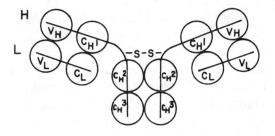

Fig. 1.3 Basic structure of immunoglobulin chains and domains, and the major proteolytic fragments that are obtained, indicated by the heavy contours. After Gall and D'Eustachio [126].

1.2.2 Location of the combining site *Confirmation of Fv.*

Direct proof that the combining site is entirely formed from the variable region has been obtained by Inbar *et al.*, [22]. The mouse IgA protein, produced by myeloma MOPC 315, with high affinity for nitrophenyl ligands (e.g. dimitrophenyl), was used. The Fab' fragment was digested with pepsin at pH 3.7 and fractionated on a Dnp-lysine-Sepharose column. The bound material was eluted specifically with Dnp-glycine and fractionated on Sephadex G-75 column. The major fraction, named Fv, has a molecular weight of 25 000 daltons and was shown to be composed of two polypeptide chains of approximately 12 000 daltons, which were separated on DEAE-cellulose in 8M-urea. One chain was shown to have the same N-terminal sequence as the intact heavy chain, whereas the other chain was blocked as is the light chain. It was concluded that Fv, which retains full binding capacity, is formed by V_L and V_H domains (Fig. 1.3). Recently another Fv was obtained also from mouse myeloma protein which is produced by the mouse myeloma XRPC 25 [23].

Further experiments demonstrated that V_L and V_H dissociated in 8M-urea can reassociate to yield an active Fv [24]. Moreover, Fv or the isolated chains were reduced in 8M-urea to open the intrachain disulfide bond and to yield a completely unfolded chain. Reoxidation of this disulfide, followed by renaturation and association of V_L and V_H, again yielded a fully active Fv [25]. These observations demonstrate that the antibody combining site is confined to the V domains and that the primary sequence of the V region solely determines the folding and specificity of the site. Hence the problem concerning the architecture of the site and the generation of its diverse specificity is confined to a sequence of 100–120 residues.

1.2.3 The size of the combining site

As mentioned earlier, the size of the antibody combining site can be estimated from the size of haptens or of antigen fragments of increasing size which inhibit the binding of the antigen by the specific antibody. This approach was used in the study of antibodies to polysaccharides, oligopeptides, or protein fragments, and demonstrates that the antigenic determinants regardless of their chemical nature, have a similar maximal size (Table 1.1). It appears that the maximum size of antigenic determinants that the antibody can accommodate is vey similar to the maximum size of substrates or inhibitors which fit the active site of enzymes. The site of lysozyme will accommodate a hexasaccharide [9] as will the binding sites of antibodies to saccharides. Similarly the binding site of papain will accommodate 6 amino acids which are similar or slightly larger than the active site of antibodies to peptidyl proteins [14].

Table 1.1 Sizes of antigenic determinants

Immunogen	Determinant	Dimensions (Å)	Ref.
Dextran	isomaltohexaose	34 x 12 x 7	[26]
Denatured DNA	pentanucleotide	28 x 10 x 10	[27]
Poly-γ-D-glutamic acid	hexaglutamic acid	36 x 10 x 6	[28]
Sperm-whale metmyoglobin	C-terminal heptapeptide	15 x 11 x 9	[29]
Poly-alanyl-protein	tri- or tetra-alanine	20 x 11 x 6.5	[14]

If the analogy between enzyme–substrate and antigen–antibody

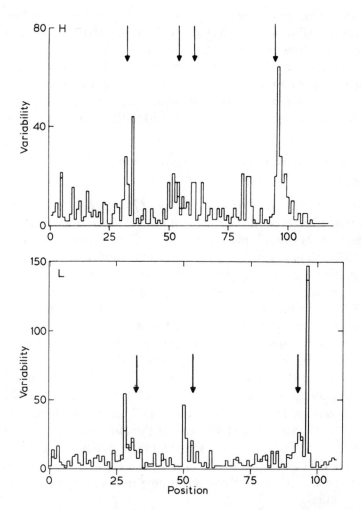

Fig. 1.4 Variability at each position in the sequence of the V region of L chain and H chain. Variability was defined as the ratio between the number of different amino acids at a given position and the frequency of the most common amino acid at that position [32, 33]. Arrows indicate the position of labeled residues obtained from various affinity labeling experiments (Table 1.2). After Wu and Kabat [32] and Kabat and Wu [33].

interaction is reflected in their structures, it implies that the architecture of the antibody combining site is similar to that of enzymes. A study of the catalytic site of lysozyme showed that it is a cleft lined by approximately 20 residues from different parts of the polypeptide chain which provides contacts with the substrate [30]. It is therefore likely that in antibodies too the combining site will be constructed from a similar number of residues contributed by each chain, and it remains to be determined which these residues are. The surprising finding was that three major stretches of amino acids at different parts of the V region were found to be involved in the combining site of all antibodies. Two approaches have been used to locate the residues which are complementary to antigen: statistical analysis of sequence variability and affinity labeling.

1.2.4 Complementary residues to antigen at the antibody combining site

1.2.4.1 Statistical analysis of variability and hypervariable segments
This approach is based on the fundamental principle that the amino acid sequence dictates the folding and the three-dimensional structure of the protein [31]. Hence, in order to learn about residues which are involved in constructing the specific site, it is necessary to compare the sequences of many V regions and to see where major differences are found. Wu and Kabat [32] made such a comparison on close to 100 different sequences of light chains, and later also on V regions of heavy chains [33]. The plot of variability versus residue position (Fig. 1.4) clearly demonstrates that in most of the positions the variability is quite limited even if sequences from different species are compared. On the other hand, three peaks of hypervariability are obvious in both L and H chains. They comprise positions 24—34, 50—56, and 89—97 of the L chains, and positions 31—35, 50—65, and 95—102 of the H chains. Since this is a statistical analysis of all available sequences, it implies that in all antibodies these regions, which together comprise about 20 residues in each chain, are the complementarity-determining residues which shape the combining site.

As will be shown later, these regions consistently appear to be implicated in the construction of the site by other independent methods. In these regions, marked variability in three-dimensional position was found by comparing Cα coordinates obtained by X-ray analysis of immunoglobulin crystals (Fig. 1.12), and these are also the regions where affinity labeling experiments were found to label particular residues (Fig. 1.4).

1.2.4.2 Affinity labeling of antibodies
This method aims at covalent binding of a ligand at the combining site of the antibody which is specific for that ligand [34]. A labeling reagent for a particular active site will have two properties. Firstly, by virtue of

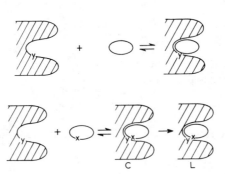

Fig. 1.5 Illustration of affinity labeling as compared with reversible binding. Top, binding of hapten by antibody. Bottom, the hapten, modified by group x, first combines reversibly with the antibody to give a complex C. While it is in the site the group x reacts to form a covalent bond with a suitable amino acid y in the site, yielding the labeled product L [35]. Reproduced with permission from *Science* [127]. Copyright 1966 by the American Association for the Advancement of Science.

its steric omplementarity to the combining site the reagent first combines reversibly with the site, and secondly by virtue of a small reactive group (x) on the reagent it reacts with one or more amino acid residues (y) in the site to form irreversible covalent bonds (Fig. 1.5). The formation of the initial reversible complex C increases the local concentration of the reagent in the site relative to its concentration in the solution. Hence, the reaction rate with residue y in the site will be greater than that with any similar residue y elsewhere on the protein. The specificity of the labeling reaction is due mainly to the increase in local concentration of the reagent in the site whereas the labeled residue need not be an un-usually reactive group [35, 36].

Singer and his colleagues introduced the diazonium labeling reagent (Fig. 1.6), and have shown that anti-benzene arsonate antibody is labeled specificially on a tyrosyl residue at the site, by diazobenzene arsonate [34]. An attractive feature of the azo-linkage formed is the possibility of reducing it with dithionite to an amino group yielding an amino-tyrosine and releasing the haptenic moiety. This was utilized by Hadler and Metzger [37] to initiate cross-linking of the L and H chains via the amino-tyrosine thus generated.

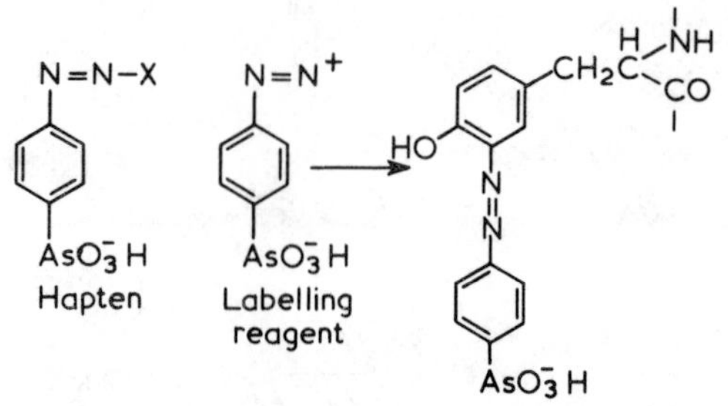

Fig. 1.6 Diazonium reagent of the hapten arsonic acid, and its reaction with a tyrosyl residue.

BADE

BADB

BADO

BADL

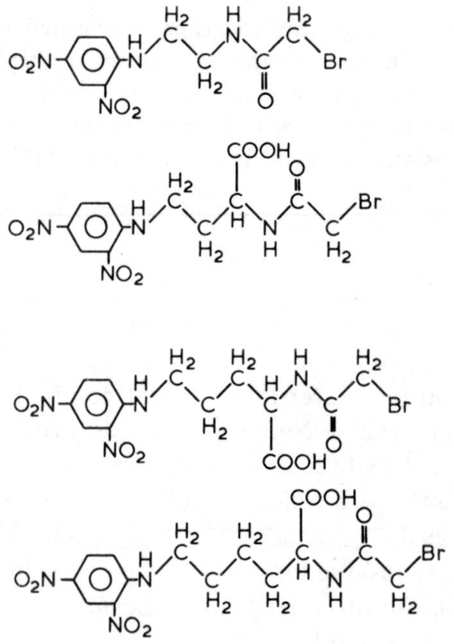

Fig. 1.7 A homologous series of bromoacetyl derivatives of Dnp-haptens [39].

Another type of affinity labeling reagent, which uses the bromoacetyl group as the functional group x, was introduced by Weinstein *et al.*, [38]. Apart from the ease of preparation of these labeling reagents and of identification of the labeled residues, it is also possible to prepare a homologous series of affinity labeling reagents (Fig. 1.7). The bromoacetyl group was attached to the terminal amino group of the hapten at

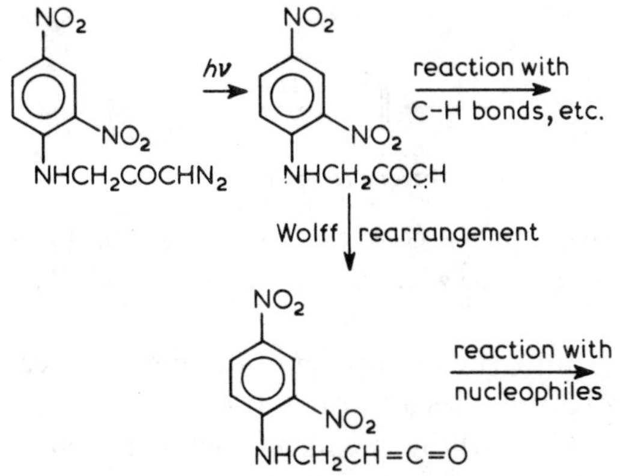

Fig. 1.8 The diazo-ketone reagent used for affinity labeling attached to the hapten Dnp-glycine. The reactions of carbene and ketene are shown [40].

systematically increasing distances from the major haptenic moiety [39], thus permitting the labeling of residues at various distances in the combining site.

Converse and Richards [40] used a diazo-ketone group attached to a dinitrophenyl hapten (Fig. 1.8). The unique feature of this type of reagent is that it is unreactive until irradiated with ultraviolet light (300 nm), so that the non-covalent and covalent binding can be separated into two steps, and excess reagent removed by gel filtration. Upon photolysis diazo-ketones form keto-carbene derivatives (Fig. 1.8) which have the ability to insert into C—H bonds. However, it appears that the carbene rapidly rearranges by a Wolff type rearrangement to the less reactive ketene which is capable of acylating nucleophiles and its range of specificity is not as wide as was desirable. At least 50 per cent of the labeling was due to reaction with the ketene.

A major point of criticism of all these reagents was that the functional group x is an additional 'tail' to the haptenic determinant and is not part of the determinant itself, thus it may label only peripherally to the site. Also only a limited number of amino acids with nucleophilic groups are susceptible to attack by these reagents. It was stressed that while, in the case of enzymes, we are constrained by selecting the reagent which fits the enzyme active site, in antibodies we can select the protein to fit the reagent by immunizing animals with haptens so designed. To overcome the disadvantages encountered with previously described reagents, Fleet *et al.*, [41, 42] introduced the aromatic azide reagent (Fig. 1.9) which

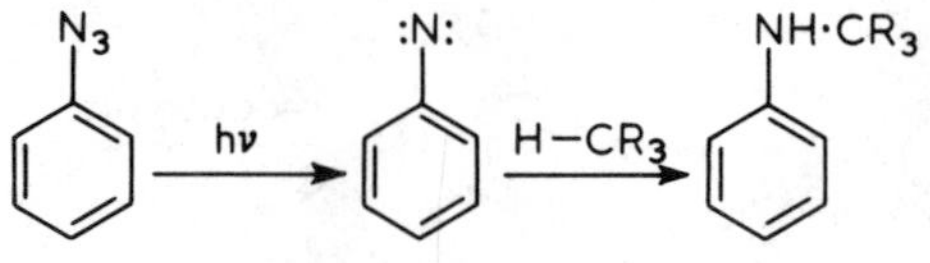

Nitrene reagent

Fig. 1.9 The aromatic azide reagent. On exposure to light the azide will decompose to give a nitrene which is able to insert into C–H bonds [41].

has the following properties. It is chemically inert and can be used as a hapten to prepare antibodies against it. It can generate photochemically (by irradiation at 400 nm) the chemically active nitrene at the combining site; and the reactive nitrene, unlike carbene, does not rearrange intramolecularly to a less reactive species and can insert into any C–H bond. Although this reagent seems to be ideal to label the protein residues which are in contact with the ligand, great difficulties were encountered in the identification of the labeled residue and the isolation of modified peptides.

It is not possible in all cases to say whether affinity labeling reagents will label contact residues in the combining site itself or residues in its close vicinity. Nevertheless, the results obtained with antibodies and myeloma proteins from different species and having different specificities were surprising in their consistency. They can be summarized as follows:

(1) Both light and heavy chains were labeled, and in most cases the ratio between the labeled chains, H/L, was around 2. With homogeneous myeloma proteins it was possible to design reagents where either L chain or H chain was labeled. Particularly interesting information was obtained with the homologous bromoacetyl derivatives of Dnp ligands and protein 315 which bind dinitrophenyl ligands. It was shown that BADE (Fig. 1.7) labeled exclusively tyrosine 34 on L chain whereas BADL labeled lysine 54 on H chain [43, 44]. This finding suggested that the distance between these two residues should be the difference in length of the two reagents. A bifunctional reagent Dnp-NH(CH$_2$)$_2$CH(HNCOCH$_2$Br) CO(NH$_2$ COCH$_2$ Br indeed cross-linked H and L of protein 315 and the label was divided equally between lysine and tyrosine [45]. These experiments again demonstrated that both chains contribute contact residues to the antibody combining site.

(2) Affinity labeling of antibodies from different species with 5 different reagents showed that the labeled residues were confined to three major stretches of the N-terminal variable region, around positions 30,

Table 1.2 Affinity-labeled residues and their location in antibodies

Antibody protein	Reagent	Labeled residue	Chain	Ref.
Pig anti-Dnp	MNBD	Tyr 33	H	[46]
Pig anti-Dnp	MNBD	Tyr 33, Tyr 93	L	[47]
Guinea pig anti-Dnp	MNBD	Tyr 33, Tyr 60, and Tyr (99–119)	H	[48]
Guinea pig anti-arsanyl	BAAT	Lys 59	H	[49]
Mouse anti-Dnp	MNBD	Tyr 86	L	[50]
Rabbit anti-Dnp	MNBD	Tyr 90	H	[50]
Rabbit anti-Dnp	NAP-Lys	(29–34), (95–114), (50–57)	H	[51]
Rabbit anti-NAP	NAP-Lys	(29–34), (95–114), (50–57)	H	[51]
Rabbit anti-NAP	NAP-Lys	Cys 92, Ala 93	H	[42]
315 anti-Dnp	MNBD	Tyr 34	L	[52]
315 anti-Dnp	BADE	Tyr 34	L	[44]
315 anti-Dnp	BADL	Lys 54	H	[44]
460 anti-Dnp	BADE	Lys 54	L	[44]
460 anti-Dnp	$DnpN_3$	Lys 54	L	[53]
460 anti-Dnp	Dnp-AD	Lys 54	L	[53]
TEPC 15 anti- phosphorylcholine	DPPC	Tyr 32, Tyr 92	L	[54, 55]
HOPC 8 anti- phosphorylcholine	DPPC	Tyr 32, Tyr 92	L	[54]

55, and 90 in either L chain or H chain (Table 1.2). These positions
indeed fall within the hypervariable segments previously described, and
Fig. 1.4 illustrates the overlap between affinity labeling and hypervariability. The hypervariable segments comprise approximately
20 per cent of the residues in the variable regions. If affinity labeling
reagents were free to combine with residues anywhere in the variable
region, then the restriction of labeling to only a few of the residues
would be very unlikely. Rather it would seem likely that residues in these
restricted segments of the variable region actually constitute the specific
combining site and are thus available for consistent labeling.

These results together with the identification of hypervariable regions
strongly suggest that the combining site is constructed by three small
stretches of the variable region of each chain. The important conclusion
is that such an arrangement is general for all antibodies regardless of
their specificity. Hence it seems that about 80 per cent of the sequence
in the V region may be considered as a framework which is not involved

in specificity, and only the hypervariable regions, which are also being affinity labeled, contribute to the combining site itself. That this is indeed the case was demonstrated by X-ray crystallographic studies of various immunoglobulin fragments, some of them with known binding activity.

1.3 THREE-DIMENSIONAL STRUCTURE OF IMMUNOGLOBULINS

1.3.1 The immunoglobulin fold

Recent progress in four different laboratories has provided high-resolution analyses of four immunoglobulin fragments of human and mouse origin (Table 1.3). The high-resolution X-ray studies have been sufficiently de-

Table 1.3 High-resolution immunogloulin structure from X-ray analysis of crystals

Protein	Species	Molecular composition	Resolution	Bound ligand	Ref.
Mcg	Human	L dimer	2.3 Å	Dnp ligands and other molecules	[56]
New	Human	Fab$'$	2.8 Å	Vitamin K_1OH	[57]
REI	Human	V_L dimer	2.8 Å	Dnp ligands	[58]
McPC603	Mouse	Fab$'$	3.1 Å	phosphorylcholine	[59]

tailed to show the folding of the polypeptide chain, the distribution of side-chains, and the combining site with its ligand. The overall dimensions of Fab$'$ are 80 x 50 x 40 Å and those of one domain are 40 x 25 x 25 Å. The results lead to several generalizations about the structure of immuno-globulin domains and about the combining site. It appears that all im-munoglobulin domains have a very similar tertiary structure based on what is called the 'immunoglobulin fold' [60, 61]. The predominant structural feature of the domain is the antiparallel β-pleated sheet. The domain is composed of two layers held like a sandwich by the single disulfide bond and enclosing a hydrophobic interior. Each layer contains either three or four straight antiparallel segments. These segments are connected by loops or bends of various lengths (Fig. 1.10). The V and C domains differ in the relative positions of the two layers mentioned above. Thus the three-segment layer is pointing either outside or inside, to meet the other chain, in the C and V domains respectively. In the C domains the four-chain layers face each other and form a large surface of interaction.

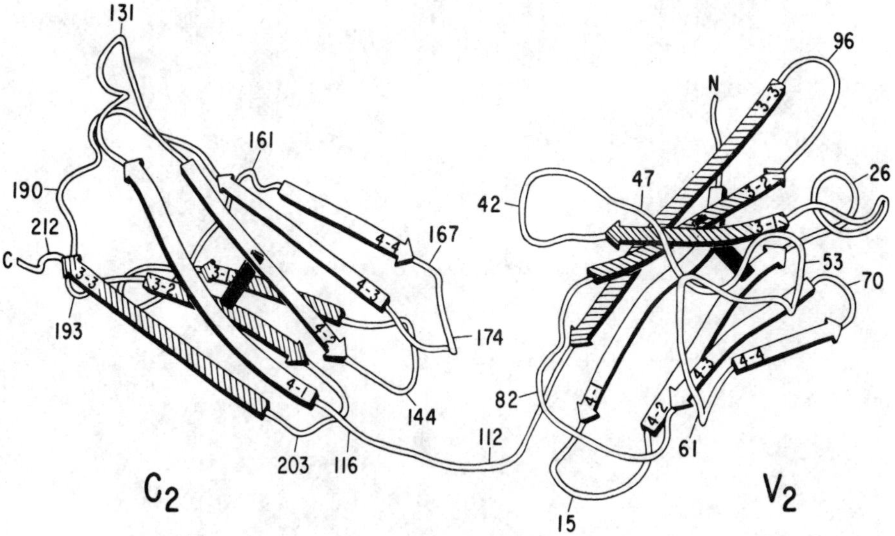

Fig. 1.10 Schematic drawing of one monomer of L chain as obtained from X-ray analysis of L dimer crystals [62]. Arrows are superimposed on segments participating in antiparallel β-pleated sheets. Three-chain layers are indicated by hatched arrows and four-chain layers by white arrows. Hypervariable residues are in some of the loops connecting the β-pleated sheets. The other chains are facing this one from above. Positions of representative residues are numbered. Reproduced with permission.

In the V domains the three-chain layers face each other and allow the formation of a combining site, by bringing together the hypervariable segments (Fig. 1.11).

1.3.2 Hypervariable and framework residues

Comparing the V domains of human or mouse origin it is observed that the relative positions of framework (i.e. non-hypervariable) residues are very similar. All the residues in the β-sheet segments of the two layers are framework residues. The hypervariable segments are in some of the loops which link these segments. The three hypervariable loops of both chains are clustered together at the amino terminus of the domain to form the area of the combining site (Fig. 1.11). The structural aspect of the concept of framework and hypervariable residues is best illustrated in Fig. 1.12. It is seen that all framework residues of V_L and V_H have their α-carbon in the same position (within $1-2$ Å) in space, relative to the domain framework. On the other hand, hypervariable residues differ

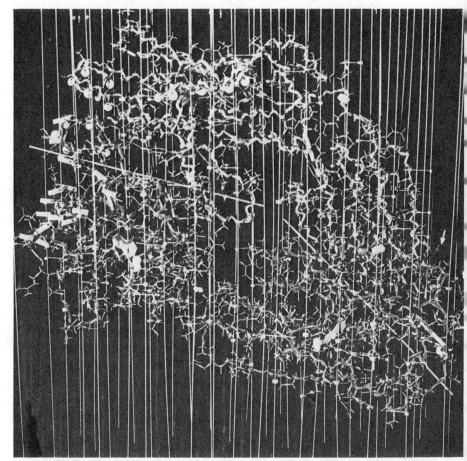

Fig. 1.11 View of the 2.8 Å resolution model of Fab' New [63]. V (left) and
C (right) domains are separated by the switch region. Rods indicate the appro-
ximate 2-fold axes of symmetry. The labels at the left of the model indicate
hypervariable positions of L (round labels) and H (rectangular labels). Reproduced
with permission.

remarkably (5–10 Å) in the position of their α-carbon in different
domains [64]. When this plot is compared with the hypervariability and
affinity labeling plot (Fig. 1.4) the similarity is obvious. In view of this
structural invariance the immunoglobulin V region may be regarded as
consisting of a rigid framework to which are attached the hypervariable
loops.

Keeping in mind the diversity of antibodies, this similarity in the
three-dimensional folding of the V domains illustrates one of the most

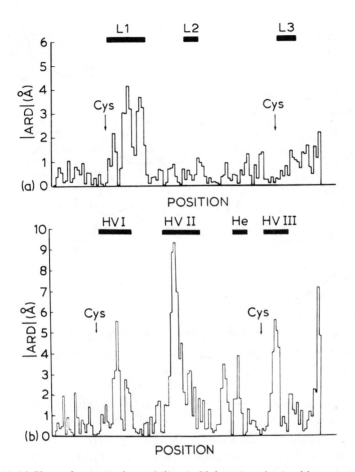

Fig. 1.12 Plots of structural variability in V domains obtained by comparing the α-carbon backbones of different domains [64]. |ARD| gives the variation in position of individual residues relative to the domain framework. Top, human V_L dimer, REI, *vs.* mouse McPC603 V_L domains. Bottom, mouse McPC603 V_L *vs.* V_H domains. L1, L2, L3 or HVI, HVII, HVIII, hypervariable segments. He specificies the position of a fourth hypervariable segment in residues 81–85 of human H chain which is external to the combining site. Reproduced with permission.

exciting 'experiments of Nature' namely *the generation of so many different combining sites on the basis of one general structure.* This idea can be tested by model building of Fv structures whose X-ray analysis was not determined owing to lack of good crystals. Padlan *et al.,* [65] built the Fv structure of protein 315 which binds Dnp ligands on the

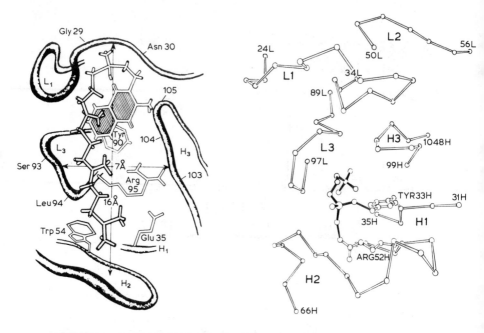

Fig. 1.13 Binding site of Fab' New [60] with vitamin K_1 OH (left), and of McPC603 [61] with phosphorylcholine (right). Hypervariable segments are indicated. Reproduction with permission.

basis of the coordinates of protein 603 which binds phosphorylcholine. In this model, several features of the binding site of protein 315 became apparent. The middle of the hypervariable surface was dominated by a rather pronounced cavity surrounded by a high density of adjacent aromatic side-chains. The results of affinity labeling by the various reagents, including the cross-linking of Lys 54 H and Tyr 33 L, are compatible with the positions of these residues in the model. Many other parameters of binding of various Dnp ligands [66, 66a] to the protein 315 are in accord with the detailed model.

1.3.3 The combining site

As discussed above the hypervariable regions of both V_L and V_H were postulated to specify the structure of the antigen binding site. Indeed the three-dimensional structure derived from X-ray analysis of Fab-hapten crystals shows that the particular association of V_L and V_H brings the hypervariable regions together to form a rather extensive and continuous antigen-binding surface (Fig. 1.13). The combining site region in Fab of protein *New* occupies an area of about 20 x 25 Å. The region that comes into contact with the hapten is relatively flat with protruding side-chains and with a central channel or pocket 15 Å wide, and with a relatively shallow depth of about 6 Å. The site is delineated by residues 27–30 and 90–95 of L chain and 30–33, 55–65, and 102–107 of H chain. By contrast the hapten binding site of protein McPC 603 is a pronounced cavity 15 Å wide, 20 Å long, and 12 Å deep whose walls are lined exclusively with hypervariable residues. It is lined by residues 27–31 and 91–96 of the L chain and 30–35, 50–59, and 99–105 of the H chain [60, 61]. Note that in both proteins the second hypervariable region of L chain is not participating in the site. This is due to deletions and insertions as will be explained later.

The structure of the binding site clearly suggests the presence of subsites which may reside on different chains. In Fab *New* which binds a derivative vitamin K_1 (K_1OH), the menadione moiety makes close contact with L1, L3 (particularly Tyr 90), and H3. The phytyl tail makes contact with L1, L3, and H2 (Fig. 1.13). At least 10–12 residues from both V_L and V_H make contact with vitamin K_1OH. Crystallographic studies of Fab *New* with menadione indicated that this ligand binds to the same part of the active site to which the menadione moiety in vitamin K_1OH is bound. Since the affinity constant for menadione is lower than that observed with vitamin K_1 OH it was concluded that the total binding energy of Fab *New*–vitamin K_1 OH complex is derived from the contacts made by both the naphthoquimone ring and the phytyl chain [60]. In the mouse Fab 603 the hapten phosphorylcholine is bound in the interior of the cavity with the phosphate group more towards the outside. The phosphate is bound exclusively to Arg 52 and Tyr 33 of H chain by electrostatic and hydrogen bonds. The choline moiety interacts with both H and L chains mainly with H3, L3, and Glu 35 of H chain.

1.3.4 Structural basis for antibody diversity

The main conclusion from the structure discussed above is that *a large number (over 50 per cent) of the amino acids of V regions are involved in constructing the framework of the immunoglobulin fold.* All the residues in the β-pleated sheets, in the hydrophobic core between the two layers, or contributin to V_L–V_H interaction must be important for the preservation of this structure. Changes in these residues will be permissible only if the replacement is by amino acids which will preserve this function. By contrast there is no structural limitation on the nature of side-chains at hypervariable positions. They occur in regions where the peptide chains are least subject to structural constraints. The number and nature of amino acids in this region will determine the binding specificity of the antibody.

In addition to amino acid replacements, insertions and deletions also play an important role in shaping the combining site. In many cases deletions and insertions in the V region occur at hypervariable segments. For example, in Fab *New* there is a deletion of 7 residues in L2, and this hypervariable region is therefore excluded from the site. In Fab 603 there is an insertion of 6 residues in L1 which causes this loop to extend further out and screen L2 from being shared in the site.

In conclusion, it seems that, although the general size of the combining site is similar in all antibodies, the actual length, width, and depth will be affected by deletions and insertions. Contact with antigen will be determined by the nature of side-chains and backbone folding of the hypervariable loops. *Thus different antibodies may display a unique antigen-complementary site although they share a common three-dimensional structure.* This can be compared with the known enzyme structures. Models of various enzymes have shown that such sites are clefts or grooves lines with 10–20 amino acids. Just how easily the specificity of binding of small molecules by proteins can be changed was demonstrated by Hartley [67] who compared of the sites of chymotrypsin, trypsin, and elastase. These sites differ in that serine 189 in the site of chymotrypsin is changed to aspartic acid in the active site of trypsin. Such a change in one amino acid will change the specificity of the combining site from the binding of an aromatic residue to the binding of a positive residue. If valine replaces glycine at position 216, as is the case in elastase, again the binding specificity will be dramatically changed.

There is no difficulty, therefore, in envisaging how such replacements in hypervariable segments will change antibody specificity, and the number of 15–20 hypervariable residues in each chain seems enough to account for as many variants as antibody diversity demands.

1.4 Mechanism of antibody action

Antibodies generally exhibit two sets of interlinked functions: the specific binding of antigen at the Fv region of the molecule and those functions shared by all antibodies of the same class, such as binding of complement, triggering of mast cells to release histamine, and, most important, triggering lymphocytes towards differentiation and antibody synthesis or towards tolerance. The latter functions are localized in the Fc part of the molecule, and under physiological conditions come into effect only after antigen binding.

We would like to understand how antigens trigger these events, and whether a 'signal' is transmitted from the Fv to the Fc region within the antibody molecule. The structural expression of such a 'signal' will presumably be an antigen-induced conformational change in the Fab, which will affect the Fc fragment.

Allosteric effects in enzymes in which a ligand induces a conformational change distal from its binding site are well known. Attempts to demonstrate such conformational changes in antibodies have yielded equivocal conclusions, although some of these studies have indicated changes in the flexibility of the molecule [68], its sedimentation coefficient [69], or its volume [70], as a consequence of hapten binding. Other studies, using optical methods such as ultraviolet absorption, fluorescence, and circular dichroism, clearly demonstrated changes in the antibody molecule that take place upon binding of hapten, but could be interpreted as changes in the combining site or in its vicinity [71, 71]. Various aspects of this problem were recently reviewed by Metzger [73] and we will summarize mainly new data concerning IgG and IgM. The main question is whether the signalling for the non-specific functions of antibodies on Fc is merely by the aggregation of antibodies by antigen, or some conformational changes in the antibody molecule are involved. Hence it is necessary to show conformational changes under conditions where monomeric antigen—antibody interactions occur.

1.4.1 Conformational changes in IgG and complement fixation

Recently evidence for conformational changes in Fab and also in the Fc upon antigen binding was obtained from measurements of the circularly polarized fluorescence (CPL) of antibodies by Givol *et al.,* [74] and Schlessinger *et al.,* [75]. The method detects very small changes in the environment or asymmetry of tryptophan residues in protein as reflected in their CPL [76]. In the above studies the CPL of antibodies was

measured before and after the binding of a multivalent antigen like
RNase at a large antigen excess, or a large monovalent antigenic determinant like the 'loop' of lysozyme. In all cases studied, CPL changes
were observed in both Fab and intact antibody. However, the changes in
the CPL spectra of the antibodies induced by binding of antigen could
not be accounted for, even qualitatively, by changes in the Fab fragments
that take place when the latter are separated from the rest of the molecule.
In fact, the changes in CPL of the antibodies and of their Fab fragments
that take place upon antigen binding occur at different spectral regions.
This clearly points to the Fc fragment as responsible for the extra conformational changes that take place in addition to that observed in
isolated Fab fragments, upon antigen binding [75]. The Fc fragment
thus seems to undergo a change in conformation upon binding of antigen
to the Fab fragments. An important aspect of this study was that reduction
of the interchain disulfide bonds in the hinge region of the antibody
abolished the antigen-induced CPL changes due to the Fc portion but
not those due to the Fab [74]. This is of interest since reduction of these
disulfide bonds, has no effect on antigen binding but does abolish
almost all complement binding [77] or C1 binding [78] by antigen–
antibody complexes. The site of complement binding is in the C_H2
domain of Fc [21] which is approximately 80–100 Å from the combining site. It is well known that small haptens would not induce complement fixation upon their binding to antibodies. Hyslop *et al.*, [79]
studied the fixation of complement by anti-Dnp antibodies in the
presence of bis-Dnp-octamethylenediamine. They fractionated the
complexes formed on Sepharose 4B column and analyzed the size of the
antibody complexes by electron microscopy. Dimers and trimers did not
fix complement whereas tetramers and pentamers of antibodies did fix
complement. In their study, the only difference between the active and
inactive complexes was the angle between the Fab arms of each IgG
molecule. The angle was less than 60° in the inactive complexes and
more than 90° in the active ones. Jaton *et al.*, [80] carried out a study
of both CPL changes and complement fixation [81] with rabbit homogeneous antibody to type III pneumococcal polysaccharide, using a
series of oligosaccharides of increasing size. They found that an
oligosaccharide of 16 units induces CPL changes in antibodies which are
attributed to Fc. However, the antibodies bound to this oligosaccharide
did not fix complement. Sedimentation analysis of these complexes
indicated the presence of dimers. On the other hand, antibodies reacting
with an oligosaccharide of 21 sugars, forming antigen–antibody complexes

higher than trimers, will bind complement.

It appears therefore that *there is an element of allostery in the antibody molecule, and antigens as well as large monovalent antigenic determinants induce conformational changes in regions distal to the combining site.* However, this may be only part of the process leading to the activation of complement binding site, and further changes are brought about by the formation of hapten antibody rings containing more than four antibody molecules.

1.4.2 Conformational changes in IgM and complement fixation

IgM is a pentamer of the basic immunoglobulin molecule and represents therefore a built-in aggregate of antibodies. According to the above discussion it is conceivable that a monovalent ligand when bound to IgM will induce complement fixation by a single IgM molecule. Brown and Koshland [82] analyzed this question using rabbit IgM antibody against the hapten phenyl-β-lactoside (Lac). They found that the hapten alone will not induce complement fixation by IgM. However mono-Lac RNase was as effective as multi-Lac RNase in inducing complement fixation. These results show that site filling by antigen is sufficient to generate changes in Fc of IgM, and that cross-linking by multifunctional antigen is not required in the IgM molecule to activate the Fc function. It was also shown that the mono-Lac RNase–IgM complex is present as monomer. These data provide strong evidence that the interaction of antigen at the Fv combining site induces a change in conformation in the distal Fc portion of the molecule.

1.4.3 Are conformational changes observed in crystals?

In the case of the two antibodies discussed above Fab 603 and *New*, no conformational change was observed after binding of the corresponding hapten to crystalline Fab [60, 61]. However, as discussed by the authors of these studies this does not rule out the possibility that such changes can occur. These Fab bind small ligands which do not interact with the total area of the combining site. Also, in the spectral analysis of CPL, previously discussed, the small hapten phosphorylcholine did not induce a change in the CPL of the antibody [75]. It seems that the structure of Fab suggests a potential for conformational change [60] resulting in a relative movement of structural subunits, similar in principle to that demonstrated for hemoglobin [83]. In the Fab′ the angle between V_L

(the variable portion of light chain) and C_L (the constant portion of light chain) domain is 100–110°, whereas between V_H (the variable portion of heavy chain) and $C_H 1$ (the first constant portion of heavy chain) the angle is 80–85° [63]. On the other hand, X-ray analysis of light chain dimer showed that one monomer behaves as the normal light chain whereas the other monomer appears to play the role of the heavy chain, having an angle of 70° between V_L and C_L [62]. This suggests that light chain and probably also heavy chain can be present in at least two different conformations. It is possible that, if a large antigenic deter- minant occupies the entire binding site, extending over the tips of both V_L and V_H, it may force a slight change in the angle between the domains in each chain. This may result in contraction or expansion of the entire Fab which will be transmitted through the rigid hinge region to the Fc portion. For this process to occur it is apparently necessary that the whole area of the binding site be occupied. It may also require that the rigidity of the hinge be maintained by its disulfide bond. A schematic representation of this possibility for conformational change in given in Fig. 1.14. Support for such a mechanism was recently obtained by X-ray analysis of the intact human immunoglobulin Kol by Colman *et al.,* [130]. Upon comparing the structure of Fab in the intact IgG with that of isolated Fab they concluded that the angle between V and C domains is different in Fab and in intact IgG. Hence they suggest that 'bending of the elbow' in the switch region between V and C domains indeed may be the basis for conformational changes in antibodies.

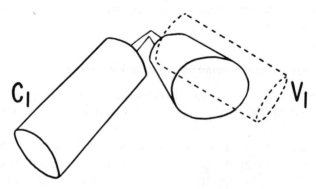

Fig. 1.14 Possibility of conformational change in L chain. Drawing of cylindrical envelopes of the V and C domains of monomer L chain in the crystal of Mcg L chain dimer [56]. The V domain is drawn also in the orientation that it would have assumed if the conformations of the two L chain monomers were identical (- - - -). The change in the angle between C and V will increase the distance between the intrachain disulfide bonds of C and V by 18 Å [129]. Reproduced with permission.

1.5 ANTIBODIES AS CELL SURFACE RECEPTORS

Cell receptor molecules, by definition, have two main functions: they recognize and bind ligands and they convey some message to the cell as a result of that binding. The diversity of antibody combining sites is so great and the mechanism of generating this diversity is probably so unique that it is unlikely that antigen recognition by cell will be mediated by another type of molecule sharing such a diversity. Rather it seems economical that lymphocyte receptors for antigen will be identical to the product later secreted by this cell, i.e. the antibody in the case of B lymphocytes. Indeed immunoglobulins were first described on the surface of lymphocytes by Sell and Gell [84]. This was demonstrated by the capacity of anti-immunoglobulin sera to activate rabbit lymphocytes *in vitro* into blast cells and to synthesize DNA.

1.5.1 Immunoglobulins on lymphocyte membrane

Stem cells of lymphocytes in mammals differentiate in the thymus into T (thymus derived) cells and in other parts of the body (probably liver, bone marrow and splee) into ('bursa-equivalent' derived) B cells. Both cell types bind antigen specifically through receptors but only B cells differentiate into plasma cells which make and secrete antibodies. It is beyond the scope of this article to discuss the nature of the receptor on T cell — whereas it is generally agreed that B cell receptors for antigens are immunoglobulins the nature of T cell receptors is still controversial. Various methods are being used to detect Ig on cell surfaces, such as fluorescent anti-immunoglobulin antibodies [85], immuno-electron microscopy [86], or labeling with ^{131}I using lactoperoxidase followed by isolation of the labeled Ig by using anti-Ig [87]. The number of Ig molecules on B lymphocytes was estimated to be between 5×10^4 to 2×10^5 [88—90]. A recent report suggests that membrane immunoglobulins are integral membrane proteins according to the definition of Singer and Nicolson [91]. This was established by testing the requirement of detergent to solubilize membrane Ig as compared with secreted Ig [92]. It is not clear if the immunoglobulin receptor on cell membrane is identical in all parts to humoral immunoglobulin, and the possibility is not excluded that it contains an extra piece which attaches it to the membrane. Although antisera prepared against L or H chain of various classes can react with cell membrane immunoglobulins, it seems that some of the Fc determinants cannot be detected on cell Ig [93],

presumably because they are buried in the membrane. Froland and
Natvig [94] prepared antibodies specific to $C_H 2$ and to $C_H 3$ regions of
the H chain and found that fluorescent anti-$C_H 2$ stains lymphocytes
whereas anti-$C_H 3$ does not. It thus seems probable that the C-terminal
domain of the antibody is embedded in the membrane whereas the Fab
portions are exposed.

Differences between the Ig class of membrane receptors and the
major Ig classes secreted by the cells are of considerable interest. Early
studies suggested that the predominant membrane Ig in mouse lymphocytes is IgM and that this is present as 7 or 8s monomer [95]. In
contrast, the major Ig in the serum is IgG. It was also demonstrated by
Pernis *et al.,* [96] that rabbit lymphocytes synthesizing IgG may have
IgM on their surface membrane. Recent progress in analysis of Ig
receptors on human lymphocytes demonstrated that IgD, an extremely
rare Ig class in serum, may be the major class of antigen cell receptor
[97]. It appears that both IgM and IgD are present simultaneously on
the same cell as true receptors, since several hours after 'stripping' of the
Ig from the cells by trypsin they reappear together on the cell [97]. It
should be made clear that, in spite of the possible coexistence of several
different classes of Ig molecules on the same cell, all the molecules on
an individual cell probably have the same specific combining sit as
analyzed by their idiotypes [98] or binding specificity [99]. This
arrangement for monospecificity is in fact an implicit feature of the clonal
selection theory of antibody responses postulated by Burnet i.e. the
same V region is expressed in receptors which have different C regions —
a finding which is related to the genetic linkage of variable region genes
of heavy chains to all heavy chain (class specific) constant region genes.
Indications were obtained [100, 101] that similar situations with regard
to IgD on cell membrane exist also in the mouse. IgD was found to be
present on more than 70 per cent of the lymphocyte. This unusual
dichotomy between Ig classes present on cell membrane and in the serum
clearly indicates that receptor IgD is not present on cell membrane to
trigger synthesis and secretion of IgD. It seems likely therefore that some
of the processes involved in cell triggering and differentiation require
IgD receptors although the cell product may be an IgG presumably with
an identical specificity. An interesting hypothesis on the role of IgM and
IgD receptor molecules in the triggering of lymphocytes was recently
published by Vitetta and Uhr [102]. They suggest that the decision
taken by lymphocytes upon antigen binding towards either tolerance,
or differentiation into antibody-producing plasma cell, is dependent on

the class of the membrane receptor. If the receptor is IgM the cell will become tolerant, whereas if it is IgD or a mixture of IgD and IgM the cell will differentiate into antibody-producing cell. The essence of their model is that IgD is a receptor designed for triggering due to its particular susceptibility to proteolysis which removes the Fab portions and unmasks another membrane protein which transmits a signal to the interior of the cell.

1.5.2 Is membrane Ig the antigen binding receptor?

The specific binding of antigen by cell receptors is the essence of the clonal selection theory [103], and such receptors were postulated as early as 1900 by Ehrlich [104] see also Greaves, M.F. in Vol 1 of this series. Nevertheless, it was not until 1967 that Naor and Sulizeano [105], using highly radioactive antigen, and by autoradiography first demonstrated that lymphocytes from non-immunized animals could bind antigen in a specific way. This study showed that 1–10 in 10 000 lymphocytes bind a specific antigen and it was also extended to measure the affinity of cell receptors and to demonstrate its similarity to the affinity of the humoral antibodies [106]. An interesting experiment demonstrating that antigen-binding cells are indeed the cells involved in the immune response was the 'antigen suicide' experiment [107]. The incubation of lympho-cytes with ^{125}I-labeled antigens with very high specific activity abolished the capacity of these cells to respond to these antigens when they are subsequently transferred into irradiated mice and challenged with the same cold antigen. The immune response to unrelated antigens was not diminished, indicating that the interference was due to specific cell destruction by irradiation from the labeled antigen.

A third line of evidence on the specificity of antigen bindings by lymphocytes was obtained by affinity chromatography of cells on solid immunoadsorbents. The incubation of lymphocytes with antigen bound to plastic particles [108], Sepharose [106], polyacrylamide [109], or nylon fibers [110] resulted in removal of the antigen-binding cells and abolished the capacity of the unbound cell population to give an immune response when transferred into irradiated mice and challenged with the corresponding antigen. There are many variations to this type of experi-ment, and the demonstration of specific unresponsiveness due to cell removal on antigen affinity columns provides convincing evidence for the specificity of cell receptors.

All these data, however, did not demonstrate that the specific receptor which binds the antigen is an immunoglobulin. Direct evidence on this point is indeed scarce, and is mostly of an indirect nature; for example the demonstration that anti-Ig can inhibit antigen binding to cells. This, together with the evidence for the existence of Ig as membrane-bound molecules, strongly suggest that the Ig is indeed the antigen receptor.

Attempts to show directly that it is surface Ig which bind antigens to cells were made by analyzing the behavior of fluorescent antigen when bound on the lymphocyte membrane. At 37° the bound antigen forms aggregates or patches and eventually accumulated in a large 'cap' at one pole of the cell [111]. Similar capping was shown to occur with anti-Ig [85]. When both antigen and anti-Ig, each labeled with a different fluorescent dye, were incubated together with lymphocytes, both cap together [112–114]. This co-capping indicates that the antigen receptor may be an immunoglobulin although other possibilities also exist. Another approach is to isolate antigen–receptor complexes from cell surface and show that they indeed contain immunoglobulin. Rolley and Marchalonis [115] used highly labeled Dnp-hemoglobin which was bound by lympho-cytes and then shed off to the medium. It was then demonstrated that the radioactive antigen coprecipitated with either anti-DNP antibodies or anti-Ig antibodies, demonstrating that the immunoglobulin receptor and antigen are in a complex. This is one of the few direct experiments to prove the co-identity of Ig and antigen receptor.

1.5.3 Immunoglobulin receptors and lymphocyte triggering by antigen

For most antigens the triggering of B cells to differentiate and produce antibodies requires, in addition to antigen binding, the cooperation of specific T cells. There are, however, some thymus-independent antigens which trigger B cells independently of T cells and therefore are better suited to analyze the requirements for B cell triggering and its relationship to the antigen receptor. It is generally agreed, now, as discussed above, that receptor immunoglobulin binds the antigen on B cells. There is, however, much less information on the consequences of this binding. Obviously we know very little about the processes of cell triggering via membrane receptors, and in lymphocytes *the central issue is whether or not the Ig receptors are involved in the signal given to the cell to differ-entiate or are they only involved in concentrating the antigen, i.e. are they 'receptors' or just 'acceptors'.* The question was recently discussed in two comprehensive reviews [116, 117], and, as in many other cases, the number of models proposed are inversely related to the information

	A	B	C
	Ag		
Activation	One signal via mitogenic site	Two signals ① via Ig receptor ② via nonspecific sites	One signal via Ig receptors
independent antigen	mitogenic	No T independence	appropriate density of repeating determinants
-dependent antigen	non mitogenic		low density of determinants

F ig. 1.15 Schematic drawing of the three hypothetical models for B lymphocyte triggering as discussed in the text. Ig, immunoglobulin receptor; Ag, antigen; M, mitogenic site which can also be triggered by non-antigenic mitogens.

available. We will therefore only summarize briefly the major opinions to show some of the different possibilities (Fig. 1.15).

The most extreme model [118] suggests that the interaction between Ig receptors and antigen does not generate any triggering signal to the cell. Such signals are generated at other surface structures (mitogenic receptors) that are present on all B cells and are not specific for antigens. The function of the Ig receptors on B cells is only to focus the antigen on to these mitogenic sites of the membrane which will then activate the cell. The T-independent antigens are therefore mitogenic by themselves whereas T-dependent antigens are not mitogenic and require other factors from T cells or the environment which will provide the mitogenic signal. In this model, only one, essentially non-specific signal triggers B lymphocytes. Such a signal may be provided either by a non-specific, polyclonal, B cell activators like lipopolysaccharide when present at high enough concentrations or on a clonal basis by certain antigens when bound by B cell Ig receptors. This model denies any function of membrane Ig as true receptors and considers them as merely selective binding instruments.

The second model [119–121] postulates two signals for activating antibody-producing cells. The first signal is a consequence of antigen binding by Ig receptors but is not in itself sufficient to trigger the lymphocyte. It is possible that this signal by itself will lead to paralysis or inactivation of the cell. Stimulation for antibody production will occur only if a second signal is being provided. This second signal may be an 'associative antibody' derived from T cells or other factors which may be non-specific. In some versions of this model, therefore, there is no truly T-independent antigen, and such antigens differ from T-dependent antigens only in quantitative aspects of the requirement for the second signal. In this model there are two versions for the first signal. One of them suggests that it must involve aggregation of Ig receptors [121] and thus may resemble other cases of lymphocyte stimulation by lectins where aggregation seems necessary, though not sufficient, for cell stimulation. The other version [119, 122] does not require the receptors to aggregate and implies conformational changes in the single receptor molecule which binds the antigen.

The third general model puts all the weight of triggering the cells, by thymus-independent antigens, on the Ig receptors on B cells [123]. It suggests that the dominant feature of B cell stimulation lies in the manner of antigenic encounter and the nature of receptor—antigen matrix that is formed. That is why only large antigens with repeating antigenic determinants can be thymus-independent. Also in such antigens the density of the determinants is critical, and according to the spacing between determinants the cell will decide between immunity, i.e. antibody production, and tolerance. Generally antigens with low determinant density will induce immunity, and those with high determinant density will induce tolerance. It is included in this model thant antigen—receptor interaction will not lead to stimulation when only a single receptor site is involved. Signals will be generated only if some lattice formation between antigen and receptors is formed. This may be accompanied by conformational changes in the Fc region of the receptor.

It is obvious that we do not yet understand the pathway of information transfer from the membrane to the nucleus leading to induction of proliferation and gene activation. The only step about Ig receptors that is understood is their binding of antigen. It is, however, possible that study of the conformation of antibody in solution and the changes induced in it by binding of antigens will also give some clues to the behavior of antibodies in cell membrane. Some of the models will have to be reconsidered when the nature of antigen-induced conformational changes in antibodies is further characterized. Othe parameters of cell triggering must await a better understanding of membrane structure and the intracellular elements such as microtubules and microfilaments that connect the membrane with the nucleus.

1.6 CONCLUSION

There is more to immunology than immunity. The unique features of the immune system are present at all levels from the genome to the intact cell. These features can briefly be described as: two genes (V and C) — one polypeptide chain; two polypeptide chains (H and L) — one combining site. Two vell types (T and B) — one antibody. The immune system is a sytem of recognition with a high power of discrimination between diverse molecular structures. It can also be studied as a model of cell triggering and differentiation. Cell—cell interaction plays an important role, in the differentiation of this system and the regulation of its activity, leading either to enhancement or suppression. In all these phenomena antibody is a key molecule. It functions either in solution or as cell surface receptor. The antibody is the sensor for antigenic signals as well as the elicited response to these signals. It may serve as an antibody, as well as an antigen (idiotype) to other antibodies. Studies on the structure and function of antibodies can be used to analyze problems concerned with molecular and cellular recognition, cell triggering, and behavior of cell receptors. The structural segmentation of the antibody molecule into domains carrying either specific or general functions may be of particular value in adapting this molecule to the above mentioned functions. Much has is now known about the detailed structure of the combining site, but less is known about structural correlates of the functions of other domains. Very little is known about several classes of Ig, such as IgD, which may be important as cell receptors, or about the molecular processes which link antigen binding and activation of the functions of the region Fc. It is very likely that further analysis of molecular parameters of antibodies in solution will also lead to a better understanding of their function as membrane receptors.

Probably the most basic unsolved problem of the immune system today is the nature of the recognition molecules present on T cells which are involved in cell-mediated immunity and in the cooperation with B cells to produce antibodies. Recent experiments [124, 125] suggest that these T cell receptors may also have Fv domains as analyzed by their idiotypic antigenic determinants. The possibility exists, however, that these Fv domains carrying the combining sites may be linked to domains different from the constant domains present in humoral antibodies. Because of the great importance of T cells in recognition of foreign tissues and in defence mechanisms, the analysis of T cell receptors presents a new challenge to investigators of the molecules of immunity.

REFERENCES

1. Schechter, I. and Berger, A. (1967), *Biochem. Biophys. Res. Comm.*, **27**, 157–162.
2. Berger, A., Schechter, I., Benderly, A. and Kurn, N. (1971), in *Peptides 1969*, North Holland, 290–309.
3. Berger, A. and Achechter, I. (1970), *Phil. Trans. Roy. Soc.* (Lond.), **B, 257**, 249–264.
4. Abramowitz, N., Schechter, I. and Berger, A. (1967), *Biochem. Biophys. Res. Comm.*, **29**, 862–867.
5. Atlas, D. (1975), *J. Mol. Biol.*, **93**, 39–53.
6. Chipman, D. and Sharon, N. (1969), *Science*, **165**, 454–465.
7. Hiromi, K. (1972), in *Protein Structure and Function* (ed. Funtasu, M., Hiromi, J., Imahori, K., Murachi, T. and Narita, K.), Vol. 2, pp. 1–46. John Wiley and Sons.
8. Cuatrecasas, P., Wilchek, M. and Anfinsen, C.B. (1968), *Science*, **162**, 1491–1493.
9. Phillips, D.C. (1966), *Sci. Amer.*, **215** (5), 78–90.
10. Blake, C.C.F., Johnson, L.N., Mair, G.A., North, A.C.T., Phillips, D.C. and Sarma, V.R. (1967), *Proc. Roy. Soc.* (Lond.), **B, 167**, 378–388.
11. Kabat, E.A. (1966), *J. Immunol.*, **97**, 1–10.
12. Kabat, E.A. (1968), *Structural Concepts in Immunology and Immunochemistry*, Holt, Rinehart and Winston.
13. Schechter, I. (1970), *Nature*, **228**, 639–641.
14. Schechter, I. (1971), *Ann. N.Y. Acad. Sci.* **190**, 394–418.
15. Landsteiner, K. (1962), *The Specificity of Serological Reactions*, p, 178, Dover Publ.
16. Schechter, I., Clerici, E. and Zazepitski, E. (1971), *Eur. J. Biochem.*, **18**, 561–572.
17. Mage, R.G. and Kabat, E.A. (1963), *Biochemistry*, **2**, 1278–1288.
18. Cisar, J., Kabat, E.A., Dorner, M.M. and Liao, J. (1975), *J. Exp. Med.*, **142**, 435–459.
19. Porter, R.R. (1973), *Defence and Recognition*. Biochemistry series one (ed. Porter). **10**, 159–198, Butterworths.

20. Edelman, G.M. (1970), *Biochemistry*, **9**, 3197–3204.
21. Reid, K.B.M. and Porter, R.R. (1975), *Contemporary Topics in Molecular Immunology*, **4**, 1–22, Plenum Press.
22. Inbar, D., Hochman, J. and Givol, D. (1972), *Proc. Nat. Acad. Sci. USA*, **69**, 2659–2662.
23. Sharon, J. and Givol, D. (1976), *Biochemistry*, **15**, 1591–1594.
24. Hochman, J., Inbar, D. and Givol, D. (1973), *Biochemistry*, **12**, 1130–1135.
25. Hochman, J., Gavish, M., Inbar, D. and Givol, D. (1976), *Biochemistry*, (In press).
26. Mage, R.G. and Kabat, E.A. (1963), *J. Immunol.*, **91**, 633–640.
27. Stollar, D., Levine, L., Lehrer, H.I. and Van Vunakis, H. (1962), *Proc. Nat. Acad. Sci. USA*, **48**, 874–880.
28. Goodman, J.W. (1969), *Immunochemistry*, **6**, 139–149.
29. Crumpton, M.J., Law, H.D. and Strong, R.C. (1970), *Biochem. J.*, **116**, 923–925.
30. North, A.C.T. and Phillips, C.D. (1969), *Prog. Biophys. Molec. Biol.*, **19**, 1–32.
31. Anfinsen, C.B. (1973), *Science*, **181**, 223–230.
32. Wu, T.T. and Kabar, E.A. (1970), *J. Exp. Med.*, **132**, 211–250.
33. Kabat, E.A. and Wu, T.T. (1971), *Ann. N.Y. Acad. Sci.*, **190**, 382–391.
34. Wofsy, L., Metzger, H. and Singer, S.J. (1962), *Biochemistry*, **1**, 1031–1038.
35. Singer, S.J. (1967), *Adv. Prot. Chem.*, **22**, 1–54.
36. Givol, D. (1974), *Essays in Biochemistry*, **10**, 73–109.
37. Hadler, N. and Metzger, H. (1971), *Proc. Nat. Acad. Sci. USA*, **68**, 1421–1424.
38. Weinstein, Y., Wilchek, M. and Givol, D. (1969), *Biochem. Biophys. Res. Comm.*, **35**, 694–701.
39. Strausbauch, P.H., Weinstein, Y., Wilchek, M., Shaltiel, S. and Givol, D. (1971), *Biochemistry*, **10**, 4392–4348.
40. Converse, C.A. and Richards, F.F. (1969), *Biochemistry*, **8**, 4431–4436.
41. Fleet, C.W.J., Knowles, J.R. and Porter, R.R. (1969), *Nature*, **224**, 511–512.
42. Fleet, G.W.J., Knowles, J.R. and Porter, R.R. (1972), *Biochem. J.*, **128**, 499–508.
43. Haimovich, J., Givol, D. and Eisen, H.N. (1970), *Proc. Nat. Acad. Sci. USA*, **67**, 1656–1661.
44. Haimovich, J., Eisen, H.N., Hurwitz, E. and Givol, D. (1972), *Biochemistry*, **11**, 2389–2398.
45. Givol, D., Strausbauch, P.H., Hurwitz, E., Wilchek, M., Haimovich, J. and Eisen, H.N. (1971), *Biochemistry*, **10**, 3461–3466.
46. Franek, F. (1973), *Eur. J. Biochem.*, **83**, 59–66.
47. Franek, F. (1971), *Eur. J. Biochem.*, **19**, 176–183.
48. Ray, A. and Cebra, J.J. (1972), *Biochemistry*, **11**, 176–183.
49. Koo, P.H. and Cebra, J.J. (1974), *Biochemistry*, **13**, 184–195.
50. Thorpe, N.O. and Singer, S.J. (1969), *Biochemistry*, **8**, 4523–4534.
51. Fisher, C.E. and Press, E.M. (1974), *Biochem. J.*, **139**, 135–149.
52. Goetzl, E.J. and Metzger, H. (1970), *Biochemistry*, **9**, 3862–3871.

53. Hew, C.L., Lifter, J., Hoshioka, M., Richards, F.F. and Konigsberg, W.H. (1973), *Biochemistry*, 12, 4685—4689.
54. Chesebro, B., Hadler, N. and Metzger, H. (1973), in *Specific Receptors of Antibodies, Antigens and Cells*, (ed. Pressman, Tomassi, Grossberg and Rose), pp. 205—217, Karger Publ.
55. Chesebro, B. and Metzger, H. (1972), *Biochemistry*, 11, 766—771.
56. Edmundson, A.B., Ely, K.R., Girling, R.L., Abola, E.E., Schiffer, M., Westholm, F.A., Fausch, M.D. and Dentsch, H.F. (1974), *Biochemistry*, 13, 3816—3827.
57. Poljak, R.J., Amzel, L.M., Chen, B.L., Phizackereley, R.P. and Saul, F. (1974), *Proc. Nat. Acad. Sci., USA*, 71, 3440—3444.
58. Epp. O., Lattman, E., Schiffer, M., Huber, R. and Palm, W. (1975), *Biochemistry*, 14, 4943—4952.
59. Segal, D.M., Padlan, E.A., Cohen, G.H., Rudikoff, S., Potter, M. and Davies, D.R. (1974), *Proc. Nat. Acad. Sci. USA*, 71, 4298—4302.
60. Poljak, R.J. (1975), *Adv. Immunol.*, 21, 1—33.
61. Davies, D.R., Padlan, E.A. and Segal, D.M. (1975), *Ann. Rev. Biochem.*, 44, 639—667.
62. Schiffer, M., Girling, R.L., Ely, K.R. and Edmundson, A.B. (1973), *Biochemistry*, 12, 4620—4631.
63. Poljak, R.J., Amzel, L.M., Avey, H.P., Chen, B.L., Phizackerly, R.P. and Saul, F. (1973), *Proc. Nat. Acad. Sci. USA*, 71, 3305—3310.
64. Padlan, E.A. and Davies, D.R. (1975), *Proc. Nat. Acad. Sci. USA*, 72, 818—923.
65. Padlan, E.A., Davies, D.R., Pecht, I., Givol, D. and Wright, C.W. (1976), *Cold Spring Harbor Symp.*, 41, (in press).
66. Haselkorn, D., Freedman, S., Givol, D. and Pecht, I. (1974), *Biochemistry*, 10,
66a. Dwek, R.A. *Contemporary topics in molecular Immunology VI*, (in press). 2210—2222.
67. Hartley, B.S. (1970), *Phil. Trans. Roy. Soc.* (Lond.), B. 257, 77—87.
68. Tumerman, L.A., Nezlin, R.S. and Zagyansky, Y.D. (1972), *FEBS Lett.*, 19, 290—292.
69. Warner, C. and Shumaker, V. (1970), *Biochemistry*, 9, 451—458.
70. Pilz, I., Kratky, O., Licht, A. and Sela, M. (1973), *Biochemistry*, 12, 4998—5005.
71. Pollet, R., Edelhoch, H., Rudikoff, S. and Potter, M. (1974), *J. Biol. Chem.*, 249, 5188—5194.
72. Holowka, D.A., Strosberg, A.D., Kimball, J.W., Haber, E. and Cathou, R.E. (1972), *Proc. Nat. Acad. Sci. USA*, 69, 3399—3403.
73. Metzger, H. (1974), *Adv. Immunol.*, 18, 169—207.
74. Givol, D., Pecht, I., Hochman, J., Schlessinger, J. and Steinberg, I.Z. (1974), *Progress in Immunology II*, Vol. I, 39—48, North-Holland, Amsterdam.
75. Schlessinger, J., Steinberg, J.Z., Givol, D., Hochman, J. and Pecht, I. (1975), *Proc. Nat. Acad. Sci. USA*, 72, 2775—2779.
76. Steinberg, J.Z. (1975), in *Concepts in Biochemical Fluorescence* (ed. Chem., R. and Edelhoch, H.), Marcel Dekker, New York, in press.

77. Schur, P.H. and Christian, G.D. (1964), *J. Exp. Med.* **120**, 531–545.
78. Press, E.M. (1975), *Biochem. J.*, **149**, 285–288.
79. Hyslop, N.E., Dourmashkin, R.P., Green, N.M. and Porter, R.P. (1970), *J. Exp. Med.*, **131**, 783–802.
80. Jaton, J.C., Huser, H., Braun, D., Givol, D., Pecht, I. and Schlessinger, J. (1975), *Biochemistry*, **14**, 5312–5315.
81. Jaton, J.C., Huser, H., Riesen, W.F., Schlessinger, J. and Givol, D. (1976), *J. Immunol.* , (in press).
82. Brown, J.C. and Koshland, M.E. (1975), *Proc. Nat. Acad. Sci. USA*, **72**, 5111–5115.
83. Perutz, M.F. (1970), *Nature*, **228**, 726–734.
84. Sell, S. and Gell, P.G.A. (1965), *J. Exp. Med.*, **122**, 423–439.
85. Raff, M.C., Sternberg, M. and Taylor, R. (1970), *Nature*, **225**, 553–554.
86. Hammerling, U. and Rajewsky, K. (1971), *Eur. J. Immunol.*, **1**, 447–452.
87. Marchalonis, J.J., Cone, R.E. and Santer, V. (1971), *Biochem. J.*, **124**, 921–927.
88. Klein, E., Eskeland, J., Inoue, M., Strom, R. and Johansson, B. (1970), *Exp. Cell. Res.*, **62**, 133–148.
89. Lerner, R.A., McConahey, P., Jansen, J. and Dixon, F. (1972), *J. Exp. Med.*, **135**, 136–149.
90. Rabellino, E. and Grey, H.M. (1971), *J. Immunol.*, **106**, 1418–1420.
91. Singer, S.J. and Nicolson, G.L. (1972), *Science*, **175**, 720–731.
92. Melcher, U., Eidels, L. and Uhr, J.W. (1975), *Nature*, **258**, 434–435.
93. Fu, S.H. and Kunkel, H.G. (1974), *J. Exp. Med.*, **140**, 895–903.
94. Froland, S.S. and Natvig, J.B. (1972), *J. Exp. Med.*, **136**, 409–414.
95. Vitetta, E.S., Baur, S. and Uhr, J.W. (1971), *J. Exp. Med.*, **134**, 242–264.
96. Pernis, B., Forni, L. and Amante, L. (1971), *Ann. N.Y. Acad. Sci.*, **190**, 420–429.
97. Rowe, D.S., Hug, K., Forni, L. and Pernis, B. (1973), *J. Exp. Med.*, **138**, 965–972.
98. Fu, S.H., Winchester, R.J. and Kunkel, H.G. (1975), *J. Immunol.*, **114**, 250–252.
99. Pernis, B., Bronet, J.C. and Seligman, M. (1974), *Eur. J. Immunol.*, **4**, 776–784.
100. Abney, R.E. and Parkhouse, R.M.E., (1974), *Nature*, **252**, 600–602.
101. Melcher, U., Vitetta, E.S., McWilliams, M., Phillips-Quagliata, J., Lamm, M. and Uhr, J.W. (1974), *J. Exp. Med.*, **140**, 1427–1431.
102. Vitetta, E.S. and Uhr, J.W. (1975), *Science*, **189**, 964–969.
103. Burnet, F.M. (1959), *The Clonal Selection Theory of Acquired Immunity*, Vanderbilt Univ. Press, Nashville, Tenn.
104. Ehrlich, P. (1900), *Proc. Roy. Soc.* (Lond.), **66**, 424–
105. Naor, D. and Sulitzeanu, D. (1967), *Nature*, **214**, 687–688.
106. Davie, J.M. and Paul, W.E. (1971), *J. Exp. Med.*, **134**, 495–516.

107. Byrt, G.L. and Ada, G.L. (1969), *Nature,* **222,** 1291–1292.
108. Wigzell, A. and Andersson, B. (1969), *J. Exp. Med.,* **129,** 23–36.
109. Truffa-Bachi, P. and Wofsy, L. (1970), *Proc. Nat. Acad. Sci. USA,* **66,** 685–692.
110. Edelman, G.M., Ruttishauser, U. and Millette, C.F. (1971), *Proc. Nat. Acad. Sci. USA,* **68,** 2153–2158.
111. Raff, M.C. and DePetris, S. (1974), in *The Immune System: Genes, Receptors, Signals* (ed. Sencarz, E.F., Williamson, A.R. and Fox, C.F.), pp. 247–257, Academic Press, New York.
112. Unanue, E.R. and Karnovsky, M. (1973), *Transplant. Rev.,* **14,** 184–210.
113. Roelants, G.E., Forni, L. and Pernis, B. (1973), *J. Exp. Med.,* **137,** 1060–1077.
114. Roelants, A., Ryden, L., Hagg, B. and Loor, F. (1974), *Nature,* **247,** 106–108.
115. Rolley, R.T. and Marchalonis, J.J. (1972), *Transplantation,* **14,** 731–741.
116. Moller, G. (Ed.) (1975), *Transplantation Reviews* **23,** 1–265.
117. Couthino, A. and Moller, G. (1974), *Adv. Immunol.,* **21,** 113–236.
118. Couthino, A. and Moller, G. (1974), *Scand. J. Immunol.,* **3,** 133–147.
119. Bretscher, P.A. and Cohn, M. (1970), *Science,* **169,** 1042–1049.
120. Cohn, M. (1972), *Cell Immunol.,* **5,** 1–20.
121. Waldman, A. and Munro, A. (1975), *Transpl. Rev.,* **23,** 213–222.
122. Bretscher, P.A. (1975), *Transpl. Rev.,* **23,** 37–48.
123. Feldmann, M., Howard, J.G. and Desaymard, C. (1975), *Transpl. Rev.,* **23,** 78–97.
124. Eichman, K. and Rajewsky, K. (1975), *Eur. J. Immunol.,* **5,** 661–666.
125. Binz, H. and Wigzell, H. (1975), *J. Exp. Med.,* **142,** 1218–1240.
126. Gall, W.E. and D'Eustachio, P. (1972), *Biochemistry* **11,** 4621–4628.
127. Singer, S.J. and Doolittle, R.F. (1966), *Science* **153,** 13–25.
128. Amzel, L.M., Poljak, R.J., Saul, F., Varga, J.M. and Richards, F.M. (1974), *Proc. Nat. Acad. Sci., USA,* **71,** 1427–1430.
129. Edmundson, A.B., Ely, K.R., Abola, A.A., Schiffer, M. and Panagiotopoulos, N. (1975), *Biochemistry* **14,** 3953–3961.
130. Colman, P.M., Deisenhofer, J., Huber, R. and Palm, W. (1976), *J. Mol. Biol.,* pp. 257–282.

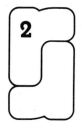

Calcium and Cell Activation

B.D. GOMPERTS
Department of Experimental Pathology,
University College Hospital Medical School,
London.

Acknowledgement

The script of this essay was read while in draft form by a number of my friends and colleagues, who provided much constructive criticism, and many ideas. In particular, I should like to acknowledge the help of Drs Clare Fewtrell, Roger Dean, Patrick Riley and Martin Raff. To Durward Lawson I am particularly grateful for the provision of electron microscope photographs, and to Dr Bob Hamil of Eli Lilly and Co. and to Dr. Julius Berger of Hoffman La Roche, I express my gratitude for the supply of the ionophores described (and many others) without which much of our work would have been impossible.

Receptors and Recognition, Series A, Volume 2

Edited by P. Cuatrecasas and M.F. Greaves

Published in 1976 by Chapman and Hall, 11 New Fetter Lane, London EC4P 4EE

2.1 INTRODUCTION

A number of the more enduring truths of science have been discovered as the result of mistakes; sometimes even from a momentary failure of vigilance coupled with uncommon perspicacity. The original discovery of a requirement for calcium in a biological process must surely be placed in this category.

In June 1882, Sidney Ringer, working at University College, London was studying the effect of the monovalent cations Na^+, K^+ and NH_4^+ on the contractility of frog heart ventricle [1]. For reasons which remain undisclosed the saline solutions were prepared from pipe water supplied by the New River Water Company. It contained 38.3 parts per million of calcium; i.e. about 1 mM. When at a later date the experiments were repeated, using saline solutions made up in distilled water (supplied by Messrs. Hopkin and Williams), it was found that the contractile function failed. Contractility could be restored by the addition of physiological concentrations (or lower) of calcium salts [2].

Hints of a wider role for calcium followed soon after with Locke's demonstration that removal of calcium could block the transmission of impulses at the neuromuscular junction in a frog sartorius preparation [3]. The realization that this effect of calcium is due to its role in controlling the secretion of a chemical messenger had to wait fifty years [4, 5].

The general role of calcium in biological activation phenomena, and more particularly, the role of calcium in stimulus–secretion coupling, has been discussed and reviewed a number of times in the past decade [6, 7]. Furthermore, the multifarious aspects of regulation and modulation of calcium mediated processes by the cyclic nucleotides has received much attention [8, 9, 10, 11]. The chief justification for stepping once again a well troden path, is the advent of the ionophores for calcium (*see* pp 68–71). By the use of these lipid soluble carrier substances of microbiological origin, it has been possible, not only to confirm much that was previously just well founded supposition, but also to extend our understanding of the many endogenous and exogenous control processes to which the activation and expression of cellular activity is subject.

2.2 TWO CRITERIA FOR ACTIVATION BY CALCIUM

In the experiments of both Locke and Ringer, about 20 min were required for the effects of calcium deprivation to become manifest [2, 3], and from this observation it was probably apparent that the role of calcium was being expressed at a site removed from the immediate vicinity of the fluid bathing the preparation.

In this account my primar concern is to show that it is the movement of Ca^{2+} ions into the cell cytosol which triggers the expression of certain types of cellular activity following the application of external stimuli. In order to support this hypothesis, two experimental approaches are fundamental. The first should endeavour to demonstrate the movement of Ca^{2+} (either from internal Ca^{2+} stores, or from the external environment) into the cytosol as a direct consequence of the stimulus. The second should be to demonstrate that the normal expression of tissue activity arises as a direct consequence of the artificial introduction of calcium ions into the cell cytosol in the absence of the physiological stimulus (external or first messenger).

2.2.1 Raising the Ca^{2+} in the cytosol

The second criterion for a calcium theory of cellular activation was satisfied first. Kamada and Kinoshita [12] in Japan (1943) and Heilbrunn and Wiercinski [13] in the USA (1947) showed that of the physiological cations, only calcium when introduced into the cytoplasm of muscle fibres could induce shortening. The other cations, K^+, Na^+ and Mg^{2+} were without effect. Whilst it was Heilbrunn's [14] contention that the effect of calcium was on the general colloid properties of the cell cytoplasm, and that the effects of ions on the isolated muscle proteins could lack biological relevance (a point of view contrary to that expressed by Szent Gyorgy [15, 16]), the importance of calcium as an intracellular messenger was assured. The introduction of calcium by micro-injection and by iontophoresis has since been used on a number of occasions (for example, to stimulate secretion of transmitters from the presynaptic terminals of squid nerve [17], and to elicit degranulation of mast cells [18]) in order to demonstrate a role for intracellular calcium in the activation of diverse cellular phenomena. The technique is of course limited to large cells and to tissues in which the expression of activity can be observed at the level of single cells.

2.2.2 Detection of Ca^{2+} influx in stimulated cells

It was subsequently shown that a wide variety of animal tissues require the presence of extracellular calcium in order to be able to express their activity [5,6]. In some cases, and under certain conditions, the effect of calcium deprivation is delayed, or is only apparent when drastic steps are taken to deplete the tissue of calcium (e.g. by treatment with the calcium chelating substance EGTA [19, 20, 21, 22]). In contrast, a large number of tissue reactions show an almost absolute dependence on the presence of extracellular calcium. Amongst these may be cited the release of 5-hydroxytryptamine from blood platelets [23], the antigen triggered release of histamine from mast cells [24], release in insulin from the pancreatic β-cell triggered by elevation of the glucose concentration [25, 26], the release of catecholamines from the chromaffin cells of the adrenal medulla [27, 28] and the secretion of hydrolases from the neutrophil [29] (polymorphonuclear leucocyte).

It was in the neutrophil cell that the second criterion for a calcium hypothesis of cellular activation was first satisfied clearly by Woodin and Wienecke in 1963 [29]. Certainly the neutrophil was an apt choice for a number of reasons. One rabbit, suitably stimulated, can provide up to 10^9 cells in 99 per cent homogeneity, and a 'good' animal will continue to do this over a period of weeks [30]. Neutrophils are in suspension, and this reduces (but does not abolish) the problem of distinguishing between extracellular and intracellular ions. This problem is of course central to any analysis, since the calcium activation hypothesis is dependent in the last resort on a determination of the internal calcium before, and immediately after, stimulation. Neutrophils also have problems peculiar to themselves: they respond to many external stimuli, and can do so in several ways [31]. They are capable of phagocytosis, secretion of lysosomal enzymes, and of directed movement (chemotaxis). All these functions are dependent on the presence of calcium and eventually it will be necessary to elucidate the separate roles of calcium in each one. Initially, this was a secondary consideration.

2.2.2.1 Net Calcium influx in stimulated cells

In the experiment of Woodin and Wieneke [29], the actual calcium content of the washed cells was determined by a complexometric titration method using a dye (plasmo corinth B) to determine the endpoint. They showed that when neutrophills are treated with staphylococcal leucocidin, there is a considerable accumulation of calcium

in the cells, which moreover, is still observed when the cells are maintained
at ice temperature.

2.2.2.2 Calcium dependent membrane depolarization

Another situation in which a net flux of calcium has been demonstrated
is in the chromaffin cells of the adrenal medulla. In this case, Douglas
and his colleagues were able to demonstrate a calcium dependent
depolarization of the membrane when the cells were stimulated by
application of acetylcholine [32]. It was concluded that this effect
arises from an inwardly directed calcium current, since the magnitude of
the depolarization increased with the logarithm of the calcium concentration. A calcium dependent and glucose stimulated depolarization of the
pancreatic β-cell membrane has been described, and this indicates that
the normal secretogogue triggers an influx of calcium in this tissue also
[33, 34]. A similar calcium dependent depolarization has been described in the isolated salivary glands of insects (the adult blowfly,
Calliphora erythrocephala) when these are stimulated to secrete by
application of 5–hydroxytryptamine [19]. As with the micro-injection
experiments of Kamada and Kinoshita, and of Heilbrunn and Wiercinski,
this approach has the possible disadvantage of being applicable only to
single cells, though of course, it is capable of great precision.

Usually, it is an increase in the permeability of the stimulated cells to
the radioactive β-emitter $^{45}Ca^{2+}$ which is measured, and Douglas and
Poisner had already shown that stimulation of the adrenal medulla with
acetylcholine increases $^{45}Ca^{2+}$ uptake [35].

2.2.2.3 Isotope measurements

Unfortunately, this simplest approach merely detects changes in the
$^{45}Ca^{2+}$ associated with the cells and gives no information about the
mechanisms involved. Whilst we are trying to demonstrate an increased
flux of calcium through the plasma membrane, the finding of increased
cell-associated $^{45}Ca^{2+}$ cannot by itself counter the alternative possibility
that new (hypothetical) calcium binding sites have been generated on, or
in, the cells as a consequence of stimulation. An alternative approach has
been to measure the efflux of radioactive $^{45}Ca^{2+}$ from cells which have
been previously exposed to the isotope at very high specific activity
[21, 36, 37, 38, 39, 40]. This is presumed to enter the cells and exchange with the internal calcium. The cells are then washed free of
external $^{45}Ca^{2+}$ before applying the specific stimulus. An enhanced
efflux of radiocalcium from the triggered cells exchanging with extra-

cellular (unlabelled) calcium is commonly seen in spite of the fact that the net flux of calcium is directed inwards down the steep concentration gradient (approx 10^{-3} M external; approx 10^{-7} M internal). In many ways this may be a better approach to the measurement of solute fluxes, but for some ions it can be invalidated by the nature of the secretory products. For example, the secretory granules of the chomaffin cells of the adrenal medulla, and of mast cells contain polyanionic macromolecules (chromo-granins [41] and heparin [42]) which can bind the radiocalcium and release it as a consequence, not of the initial stimulation and cell trigger-ing, but as a result of the ultimate secretory event. In mast cells, the presence of heparin containing granules may also confuse the results of calcium influx experiments. In other situations, such as the β-cells of the endocrine pancreas, the plasma membrane may even respond to the external stimulus by a reduction in calcium permeability [38, 39]; the calcium for activation is then provided from stores within the cells.

When dealing with actual tissues (most notably smooth muscle) as opposed to cell suspensions, there arises the additional problem of accounting for the radiocalcium present in the vascular spaces and the interstices between the cells, which can easily contain such overwhelming quantities of the tracer that fluctuations in the cellular calcium become insignificant. This problem has now been largely overcome following the realization that lanthanum ions (La^{3+}), due to their greater charge density have a higher affinity than calcium for any anionic site which binds calcium [43]. La^{3+} can thus displace Ca^{2+} from all binding sites on the surface of cells, including the binding sites associated with the calcium transport pathways. Tissues can then be washed thoroughly over extended periods of time, thereby removing the external adsorbed calcium while the internal calcium remains locked in. The proceedure has been used successfully on smooth muscle [44, 45], and is currently being applied by us to study $^{45}Ca^{2+}$ fluxes in serum stimulated fibroblasts in mono-layer culture (Damlugi, Riley and Gomperts, unpublished observations).

2.2.2.4 Indicator measurements
Another approach to the study of calcium fluxes, and the measurement of the calcium concentration in the cytosol has been to make cells self-indicating. There are a number of dyes, e.g. murexide (ammonium purpurate) which respond to the presence of calcium by a shift in their absorption spectrum. In toad muscle which has been allowed to accumulate murexide, it was possible to show that there is a relatively rapid formation of the Ca-murexide complex which clearly preceeds the

development of tension, and returns to its initial value before the point
of maximum tension is reached [46]. The problem with applying the
techniques of spectrophotometry to contractile and secretory tissues is
that small absorption changes can easily be masked by much larger light
scattering changes. A much more potent approach, but applicable only
to cells of large dimensions has been to use the calcium sensitive photo-
protein aequorin as an indicator of internal calcium. In this way it is
possible to make calibrated measurements on a rapid, temporal basis of
the changes in calcium *activity* associated with the expression of the
cellular response. This approach has been extensively discussed by specialists in the field, with particular reference to the large single muscle
fibres of certain crustacea [47], and the giant axon of the squid [48].
Another technique, for making cells self-indicating, but which requires
only simple resources, is to treat cells with chlortetracycline (which binds
generally to membranes); in the presence of calcium the fluorescence of
this substance increases [49]. Chlortetracycline has been used to study
calcium fluxes in the pancreatic islets of Langerhans [50].

2.3 WHY CALCIUM?

Why is calcium (and not magnesium) used by cells as a second messenger
in triggering? Magnesium is frequently antagonistic to calcium mediated
processes [28, 51, 52, 53, 54, 55] but is required for many of the
metabolic (homeostatic) functions within the cell. A requirement for
extracellular magnesium for cell activation is rare: examples of this are,
the phagocytic function of human blood leucocytes and rabbit alveolar
macrophages [56] (in which Mg^{2+} may be replaced by Co^{2+} or Mn^{2+};
Ca^{2+} is very much less effective) and the non-T cell mediated lysis of
antibody coated sheep red blood cells [57] (in which Mg^{2+} can be
replaced by Mn^{2+} but Ca^{2+} is without effect).

Calcium and Magnesium are clearly recognized as different by cells,
and so it is unlikely that the role of these divalent cations is simply one
of charge neutralization, for in this case the two cations should be equally
efficient. On other hand, chemically, magnesium is clearly distinguished
from calcium by the precisely defined range of structures into which it
will fit [58]; in the formation of polydentate complexes, calcium is
subject to surprisingly few constraints. The coordination number for
calcium can vary from six to ten; for magnesium a rule of six is rigidly
adhered to. Similarly there is great variability in the bond angles and

distances which are acceptable for the formation of stable calcium compounds. Thus, for the metabolic good health of a cell, it is essential to exclude calcium from the cytosol, because calcium will in general displace magnesium from its complexes with the essential metabolic intermediates such as ATP which are only active as magnesium salts [59].

2.3.1 Regulation of intracellular calcium

It is apparent that if cells are to be responsive to calcium, they must have efficient means of disposing of it. The calcium dependent ATPase is a universal component of animal cell plasma membranes and even on such inert cells as the mature mammalian red cell, there is sufficient activity to expel large quantities of calcium from the cytosol in a very short period of time [60]. Furthermore, the transport ATPase is activated by intracellular calcium at micromolar levels and below. In skeletal muscle, calcium is sequestered in the sarcoplasmic reticulum, the membranes of which contain a calcium dependent ATPase, and little else in the way of active proteins [61]. The plasma membrane and mitochondria of heart muscle and other tissues also play a very active part in sequestering calcium away from the cytosol. In any consideration of the process of cellular activation by calcium, the subsequent fate of the calcium and of the pathways by which it initially enters the cytosol are of central importance.

In common with all the other intracellular binding sites for divalent cations, the affinity of the calcium pumps are higher for calcium than for magnesium, and so it is calcium which is selectively excluded from the cytosol. This does not explain why it is calcium which acts as a trigger, for not only must the signalling ion be highly mobile in order to ensure sensitivity (and calcium is satisfactory on this point) but it also has to have an active role once it gets into the cell. In the case of muscle, the early predictions of Szent Gyorgy [15, 16] have by now come to fruition and the event of calcium binding to a defined receptor site, troponin, the controlling unit of the myofibril, is now understood to be an instrument of the contractile process [62].

2.3.2 What does calcium do?

In other cell processes the picture is far from clear, and the argument is still in some ways at the stage described earlier, when Heilbrunn suggested that the effect of calcium is on the general colloid properties of the cell protoplasm [14]. Experience with isolated proteins of muscle would indicate that this is unlikely to be the case, and certainly, most cells investigated have been shown to contain extensive arrays of actin and myosin-like filament proteins which probably have a contractile function [63, 64]. They appear to be attached to the inside of the plasma membrane, and in the chromaffin cell of the adrenal medulla [65] and in the mast cell [66], they may also be joined to the membrane of the secretory granules, so it is likely that they play a role in secretion. So far however, the troponin element of muscle, which through tropomyosin has the effect of conferring sensitivity to calcium to the myosin ATPase has not been conclusively identified in non-muscle cells. Actin from non-muscle systems is able to bind troponin-tropomyosin derived from muscle, thereby becoming calcium sensitive [67], and there are reports of tropomyosins from cells [68] as diverse as the myxomycete *Physarum polycephalum* [69], as well as platelets [70] and brain [71]. Since crude actomyosins from a number of non-muscular tissues appear to have calcium regulated ATPases (e.g. brain [72], blood leucocytes [73], and myxomycetes [74]) it does seem likely that a non-muscle equivalent of muscle troponin will be discovered in due course*

The integrity of microtubules also varies *in vitro* with calcium concentration [75, 76, 77], so these too could be another possible site for calcium mediated regulation, and without doubt, other plausible calcium receptor sites will be uncovered in the course of time.

The problem here is to be found in the imprecision of the pharmacological tools at our disposal for distinguishing between the various subcellular structures. Treatment with any one of a number of drugs has been shown to prevent the calcium mediated responses of a large number of cells, (one thinks mainly of the cytochalasins, and the 'anti-mitotic' drugs colchicine, vinblastine etc.), but none of these appears to be capable of exerting just a single, uniquely directed and universal effect. To take the exocytotic (secretory) processes alone, two main ideas dominate the field. One line of thinking suggests that calcium exerts its intracellular effect on the microfilaments (muscle like elements) close to the periphery of the cell. By contracting, these draw the secretory granules towards the plasma membrane, with which they then fuse, and from which they are expelled [78, 79]. Another argument suggests that calcium acts as a cross-linking structure between opposing fixed negative

---

* *note added in proof*: this prediction would appear already to have been fulfilled with the report [280] of a calcium binding protein from adrenal medulla having close structural and physical resemblence to troponin-C.

charges on the membrane surfaces of the secretory granule and the internal surface of the plasma membrane [80]. This cross-linking function is one to which calcium, unlike magnesium is suited [58]. These arguments are not of course mutually exclusive, nor do they exclude the possibility of additional roles for calcium ions in the mechanisms of endocytosis and exocytosis. One idea which does not appear to have been expressed previously is that calcium could act to uncover nascent recognition sites on the surface of the membrane of a secretory granule, which would then seek out complementary receptors (assisted possibly by calcium mediated movements of the microfilament assembly) on the plasma membrane. Such a mechanism could ensure the specificity of fusion whereby granule membranes fuse with the plasma membrane and each other, but not with the membranes of the other intracellular organelles.

2.3.3 Microtubules and microfilaments

It is now widely thought that the microfilaments play a role in endocytotic and exocytotic processes, but a precise understanding of how this is expressed still eludes us [78, 79, 81]. The most powerful tools at hand for probing microfilament function are a class of fungal metabolites called the cytochalasins [82]. These compounds bind to certain intracellular contractile filaments which also have the property of binding heavy meromyosin [63, 83, 84, 85] and anti-actin antibodies [86] and are therefore understood to be 'actin-like'. After treatment with cytochalasin B, the motile functions of cells are interrupted reversibly [87, 88, 89]. Inhibitory actions of the cytochalasins on both endocytotic and exocytotic functions of cells have been described. For example, both pinocytosis and phagocytosis by mouse peritoneal macrophages are inhibited within a few minutes after treatment with cytochalasin B in the range $1-10$ μg/ml [81]. These changes are reversible. By contrast, colchicine which acts to depolymerize microtubules [90] does not inhibit these functions although it does prevent the directional movements of these cells. In certain systems, the cytochalasins also act to prevent the extrusion of granules which occurs in exocytosis. This has been described for mast cells [78] and thyroid [91], but there is no accord on the precise mechanism of inhibition. There are two problems to be considered. Firstly, cytochalasin B (on which most experience has been gathered to date) is known to be an inhibitor of glucose in most cells [92–97] and so it is absolutely vital to be sure that inhibition of mediator release and of degranulation does not arise as a result of metabolic deprivation. It has also been shown to inhibit thymidine transport in chinese hamster

ovary cells [98]. Secondly, in some systems cytochalasin B actually
potentiates secretion. A good example of this is to be found in the
secretion of insulin from the pancreatic β cells in which cytochalasin
(10 μg/ml) enhances both early and late phase secretion stimulated by
either glucose or by leucine, but has no effect by itself in the absence of
the external stimulus [99]. The effects of cytochalasin in this system are
once again reversed within a few minutes of its removal. The opposing
effects of cytochalasin B have even been observed for separate functions
of a single cell type. Thus it acts to inhibit particle uptake by human
neutrophils [100] and it is capable of stimulating the release of
β-glucuronidase [101, 102, 103]. In pancreatic acinar cells it has the
effect of a secretogue when applied at very low concentrations, but is an
inhibitor of carbamylcholine stimulated release of lipase and chymo-
trypsinogen when applied at higher concentrations [104]. In the latter
tissue, although cytochalasin B inhibits glucose transport, no alteration
in intracellular ATP was detected, nor was there any evidence of the
redistribution of calcium. The inhibitory phase of cytochalasin B could
be correlated with the disruption of the microfilamentous network and it
is thought unlikely that glucose deprivation has a significant role in this;
the mechanism whereby it acts as a secretogogues at low concentrations
is not adequately explained.

2.3.4 Secretion by exocytosis

The contents of secretory granules are released from cells by the mechanism of exocytosis [105]. By this process, the entire contents of the
granule are expelled without apparent damage to the cell. In secretion
from the adrenal medulla stimulated by acetylcholine (or carbamylcholine)
the molar ratios of the released materials (catecholamines, nucleotides
and proteins) are the same as they are in the chromaffin granules obtained
by careful homogenization and fractionation of the unstimulated tissue
[106, 107]. There is no evidence of loss of specific enzymes from the
cytosol (e.g. lactate dehydrogenase) and the loss of lipid from the gland
is only elevated slightly. The release of 5-hydroxytryptamine from
stimulated blood platelets is similarly accompanied by the other contents
of the secretory granules present in the intact cell, including non-
metabolic ATP and lysosomal enzymes [108, 109]. In the mast cell, it
has not been possible so far to separate the release of histamine and the
other mediators of the acute inflammatory processes from the release
of lysosomal enzymes, particularly β-glucuronidase (Dean, Fewtrell and
Gomperts, unpublished observations) and hexoseaminidase [110]. In no

case is there evidence of selective release of individual components present within a single granule. In some cells e.g. platelets, there are more than one type of granule, and these may be released selectively by different stimuli [108, 109], so the question of selective release must be examined critically in every case.

In the exocytosis process, the membrane of the secretory granule fuses with the internal surface of the plasma membrane: under the electron microscope, this appears as a pentalaminar structure (three stained lines separated by two unstained regions; the central line, which is darkest, represents the very close apposition of the two protoplasmic surfaces of the granule membrane and the plasma membrane which otherwise appear as typical trilaminar structures). The granule is exteriorized through this fused structure, which re-anneals around the cytoplasmic side of the granule to retain the integrity of the cell membrane. Ultrastructural evidence for the exocytotic process was first described in the enzyme-secreting pancreatic acinar cells [111] but evidence of membrane fusion has now been found in the stimulated cells of almost every secretory tissue which has been examined. Membrane fusion in secretion is a calcium dependent process.

2.4 THE RAT PERITONEAL MAST CELL

At this stage, I intend to focus attention mainly on a few cell types, chief among these being the rat peritoneal mast cell. The following section offers some justification for this choice.

Mast cells are present to the extent of about 2–5 per cent in the cell suspension obtained by peritoneal lavage of a rat. In many ways they offer a quite unparalled opportunity to observe at a variety of levels, the processes which result from the initial attachment of an external messenger to a receptor on the cell surface, and which lead to the ultimate expression of cellular activity. These cells may easily be obtained in homogeneous suspension by density gradient centrifugation; their only limitation is the small quantity of cells which are normally obtained: commonly, we are able to isolate about 0.25×10^6 cells per animal. By appropriate immunization one may control the identity of the superficial receptors on the cell surface. These receptors are immunoglobulin molecules of the class IgE [112, 113, 114]. They are attached to cell surface receptors (Fc receptors) which recognize the Fc portion of the IgE molecule. The antigen directed Fab portions are exposed and so such sensitized cells can be provoked to release histamine (which is very easily measured)

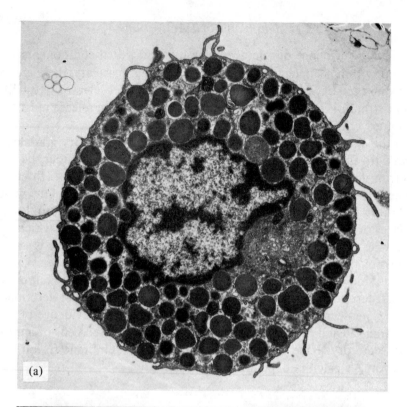

(a)

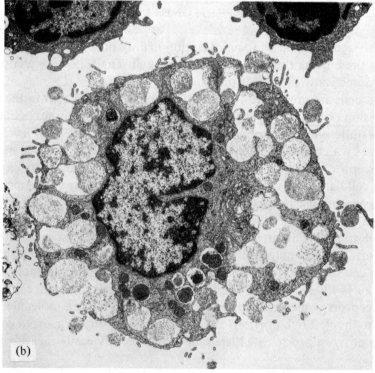

(b)

following exposure of the cells to the specific antigen [115], to anti-rat immunoglobulin [116] or to the plant lectin, concanavalin A (con A) [117]. After binding of the ligand, exocytosis takes place within a few seconds, and terminates short of total expulsion of the secretory material within a minute or so [40, 118]. The process is dependent on the presence of external calcium [24, 55]. All three of these releasing ligands are di- or multivalent. That is to say, they have two or more points of attachment and so are able to cross-link the surface IgE receptors. It is possible to prepare monovalent Fab fragments by digestion of anti-immunoglobulin with papain; Fab is capable of binding to the IgE receptors of the mast cell but it is quite incapable of stimulating release. The requirement then is for ligands that can cross link between surface IgE molecules [118–122]. With con A, anti-Ig and Fab, one may covalently attach ferritin, and still retain a high affinity for the surface receptor sites; con A-ferritin and anti-Ig-ferritin are both capable of stimulating release [122]. The ferritin molecule is electron dense ans shows up as a recognizable black spot on unstained, thin-section electron microscope photographs; by using this technique, one may follow the subsequent fate of the triggering ligands after they have bound to their receptors on the cell surface.

2.4.1 Cyclic AMP: second messenger or inhibitory modulator?

Another attraction of mast cells concerns the clearly defined effects of cyclic 3′, 5′-adenylic acid (cyclic AMP). All manipulations which have the effect of elevating the concentration of this regulating metabolite in the cytoplasm of the mast cell are inhibitory to the secretory process [40, 123–129]. The effects of manipulations intended to elevate cyclic GMP levels have been investigated more fully on human lung mast cells, and here the picture is one of augmentation of the secretory process by cyclic GMP [129]. Depression of cyclic AMP appears to have little effect

Fig. 2.1 (a) Thin section electron microscope photograph of an unreleased mast cell. The granules are seen to be compact, and contained within individual membrane bound organelles. (b) Electron microscopic appearance of a well-released mast cell following treatment with sheep anti-rat immunoglobulin. Only a single granule now appears unaltered; the rest have either been shed from the cell, or are contained within labarinthine cavities which are in contact with the extracellular space. (From Lawson, Fewtrell, Gomperts and Raff, [122].

Table 2.1 Modulation by cyclic nucleotides of three distinct effector systems of inflammation*

Agents	Antigen-induced, IgE-dependent secretion of chemical mediators from human lung tissue	Lysosomal enzyme release from polymorphonuclear leukocytes	Lymphoc mediated cytotoxic
Effect of agents increasing cyclic AMP levels			
1. Dibutyryl cyclic AMP	I	I	I
2. Phosphodiesterase inhibitors	I	I	I
3. β-adrenergic agents	I	I	I
4. Prostaglandin E$_1$	I	I	I
5. β-adrenergic agents + propranolol	NE	NE	
6. Cholera toxin	I		I
Effect of agents decreasing cyclic AMP levels			
1. α-adrenergic agents	E	NE	
2. Imidazole	E		E
3. Prostaglandin F$_{2\alpha}$	E		
Effect of cholinergic agents			
1. Acetylcholine or carbamylcholine	E	E	E
2. Carbamylcholine + atropine	NE	NE	NE
3. 8-bromo-cyclic GMP	E	E	E

*Abbreviations: I—inhibitory; E—enhancing; NE—no effect;
From Kaliner and Austen [129].

either way. Inhibition of calcium dependent functions by cyclic AMP seems to be the rule in a number of processes linked to the immune system (see Table 2.1), but this is by no means a general phenomenon.

As examples of calcium dependent, cyclic AMP inhibited processes, we may include the multiple functions of the neutrophil [130, 131, 132] and DNA synthesis and blast formation in the lymphocyte [133]; extending our view beyond the immune system, it is apparent that shape change and the secretory processes of blood platelets [134], and DNA synthesis and proliferation of some fibroblasts grown in monolayer culture [135, 136] are all subject to a similar control. Barger and Dale, working in 1910, were probably the first to show that contraction of certain types of smooth muscle could be prevented by application of sympathomimetic amines [137], and some of the effects which they noted were undoubtedly due to elevation of cyclic AMP.

In other calcium dependent secretory functions, e.g. the release of catecholamines, from the cat adrenal, cyclic AMP appears to be without effect [138] although it is generated in response to stimulation. The time course of its generation is distinctly out of phase with the secretory response, and it is thought that it may modulate catecholamine resyntheis through an effect on tyrosine hydroxylase [139]. However, treatment of pancreatic islets with theophylline or with dibutyryl cyclic AMP potentiates the release of insulin stimulated by normal secretogogues [140], and cyclic AMP has been implicated as an essential intermediate in the calcium dependent steroidogenic [144, 145], but not in the secretogogic effect of ACTH on adrenocortical cells [141]. In the latter case it appears that calcium is required [142] as an enabling factor in the ACTH stimulated formation of cyclic AMP [143]. It is cyclic AMP which is in effect the 'second messenger' [146]. In the β-cells of pancreatic islets and in blowfly salivary gland, cyclic AMP plays the role of modulator, not of second messenger which in both these cases has been shown to be calcium [39, 140]. However, the situation is more complicated than this because in glucose stimulated islets the intracellular level of cyclic AMP is elevated in a manner which is dependent on the concentration of external calcium [147, 148]; in the blowfly salivary gland it is the opposite which is observed, and calcium is antagonistic to cyclic AMP formation [149]. In the β-cell, elevation of cyclic AMP due to phosphodiesterase inhibition with theophylline, enables both glucose and leucine to stimulate insulin secretion in the absence of external calcium [140], and so it is likely that the effect of cyclic AMP in this case is to permit the release of calcium from intracellular stores. This could explain why cyclic AMP acts as a potentiator of calcium dependent insulin secretion.

The complex interrelationships which exist between the metabolism of cAMP and the availability of calcium, and the multifarious effects which these two components of the cellular activation sequence have on the ultimate expression of cellular activity have been reviewed and discussed at length [8, 9, 10, 11]. The purpose of the above remarks in the present contect is to stress the choice of the mast cell as an ideal model for investigating the processes involved in cellular activation. Here, cAMP is clearly antagonistic, an experimental finding based on 40 years of experience with substances now known to exert their effects either as activators of the adenylyl cyclases, or as inhibitors of phosphodiesterase. This unambiguous inhibitory role of cAMP holds out the possibility that by working with mast cells, we may be able to delineate with some precision, and greater chance of success, at least one role where cyclic AMP is a modulator of the secretory process.

2.5 CALCIUM INDEPENDENT PROCESSES

When cells are treated with triggering ligands, a number of processes are set in motion, some of which are clearly independent of the presence or absence of calcium. It is among these that we might expect to find clues to the cellular mechanism which regulates calcium entry.

2.5.1 Clustering, patching and capping

As was stated above, one may use ferritin labelled ligands to stimulate secretion in the mast cell, and this allows us to study the fate of the triggering ligand after attachment to its receptor sites on the cell surface. The phenomenon of patching was first described for B (immunoglobulin secreting) lymphocytes [150]. When treated with divalent anti-immunoglobulin-ferritin for a few minutes, the fixed and sectioned cell reveals the ferritin to be distributed in discete patches on the cell surface [151]. These patches are in reality two dimensional immune precipitates generated by the cross linking of the surface receptors by the divalent ligand. Extensive regions of the cell surface of a patched cell are devoid of the ferritin label. By contrast, treatment of the cells with a monovalent ligand (Fab-ferritin) reveals the unperturbed condition of the receptors which are seen to be randomly distributed. Another way of seeing the normal or unperturbed distribution of receptors is to fix the cell with glutaraldehyde, which cross links amino groups, and then to treat the fixed cell with ferritin labelled con A (con A-FT), which binds to accessible glycoside residues on the cell surface. Using this method, the con A binding sites are also seen to be randomly distributed. Patching by divalent ligands is not dependent on the presence of calcium, nor on cell metabolism. It proceeds just as rapidly in cells treated with respiratory inhibitors, and is merely slowed up, not prevented, by cooling to 4°C.

Patched B cells incubated at 37°C proceed to develop a 'cap'. This involves the directed movement of the surface precipitates (patches) to the pole of the cell furthest removed from the nucleus, from where it may be ingested. In contrast to patching, capping is an active process, and may be prevented by the application of metabolic inhibitors, and a number of drugs including colchicine, cytochalasin B, (*see* pp 53—54) and local anaesthetics.

While it has not been possible (in our hands) to use ferritin labelled antigens (or ferritin as an antigen) to stimulate mast cells, they may be successfully triggered with ferritin labelled anti-immunoglobulin or con A [122]. Thus one may in this case follow the fate of the relevant transducing receptors on the cell surfaces, and determine the degree of surface redistribution (if any) which is commensurate with release. Using this approach, it was possible to show that sensitized mast cells when incubated for 0.5 min in the presence of an optimal concentration of ferritin labelled sheep anti-rat Ig serum, bound about 6 500 molecules of anti-Ig per cell and that the maximum cluster size of aggregated receptors on the membrane surface commensurate with the triggering of the cell, contained less than ten Ig molecules [122]. Capping was never observed (though it is seen on the closely related circulating blood basophil leucocyte [152]) and pinocytosis of bound ferritin only became apparent after periods of incubation approaching 30 min. Capping and pinocytosis are therefore not necessary events, but the cross linking of a very small number of IgE molecules may be required for release.

The cross linking and consequent redistribution of the surface receptors is independent of the presence of calcium, and so it is likely that it preceeds and may even determine the subsequent course of calcium dependent events. The idea has been voiced on a number of occasions that an ion pathway could be assembled through association of monomeric subunits (the intramembrane particles possibly) as a result of aggregation of surface receptors [153, 154, 154a], and there is an excellent model for this in the excitability inducing antibiotic alamethicin which appears to aggregate into a conducting transmembrane structure in electrically polarized phospholipid bilayer model membranes [155]. There is some evidence from the mamalian red blood cell that manipulations which can be shown to cause the aggregation of the surface sialoglycoprotein also tend to aggregate the intramembraneous particles seen on freeze-fracture electron microscopy [156], and even the spectrin which is present exclusively on the cytoplasmic membrane surface [157]. In contrast, it appears that in nucleated cells, the effects of surface aggregation are not transmitted in the same way. There is no evidence that patching, or even capping of surface receptors is associated with systematic alterations in the distribution of particles in the fracture plane [158, 159], even though it is now generally believed that the transport pathways of cell membranes are associated with these structures [160]. It seems more likely that the ion channels go through the transmembrane particles, and not between them, but the precise processes of transduction, whereby external stimuli are converted into commands for cellular expression, and also the mechanism of formation and even the very nature of the calcium pathways, elude us.

2.5.2 Limited proteolysis

It has been known for some time that diisopopylphosphofluoridate (DFP) [161] and certain organophosphorus compounds [162] are effective inhibitors of antigen mediated release of histamine from mast cells. DFP reacts with specific serine residues located in the region of the active site of the serine proteinases and esterases, including chymotrypsin [163] and a role for a proteolytic step has therefore been sought for in the sequence of events leading from initial attachment of the antigen to the IgE at the cell surface to the ultimate degranulation and release of histamine [164]. By itself, the evidence for the involvement of a proteolytic step based simply on inhibition by DFP is not particularly strong, especially as it has been shown that the effect of this compound correlates rather well with its inhibitory effect on mast cell glycolysis [165] (NB: the esteratic function of at least one of the enzymes of the glycolytic pathway, phosphoglucomutase, can be inhibited by DFP [163]). Further support for the involvement of a serine proteinase has come from experiments in which exogenous chymotrypsin and other enzymes have been applied to mast cells [166, 167], and it appears that the consequent degranulation is normal in every way. It is dependent on the presence of calcium [168], it is prevented by application of metabolic inhibitors [166], and it is morphologically indistinguishable from the antigen mediated exocytosis process [169]. This effect of proteolytic enzymes in stimulating cell function is by no means restricted to mast cells, and in the case of platelets, activation by the proteolytic function of thrombin (another serine proteinase) can be counted as a normal physiological process [170, 171]. Since the phagocytic [172] and chremotactic [173] functions of neutrophils can also be prevented by the organophosphorus inhibitors the possibility exist that here too, the normal stimulus to the cell effectively reveals the activity of an endogenous nascent enzyme. The inhibited neutrophil functions could be rescued by application of certain esters, most notably acetates [174], and it is therefore thought that it is an esteratic enzyme which is activated by the phagocytic and chemotactic stimuli.

DNA synthesis in lymphocytes stimulated by mitogenic lectins [175, 176] and in cultured fibroblasts stimulated by addition of serum [177] (both calcium dependent functions) can also be prevented by treatment with inhibitors of the serine proteinases. Whilst there is absolutely no reason to think that the common mitogenic substances (e.g. concanavalin A for T lymphocytes) have a proteolytic function themselves, it does appear that they are able to activate a very limited degree of proteolysis at the cell surface since the application of a number of proteolytic enzymes to these cells can trigger off DNA synthesis and mitosis [178, 179, 180, 181]. Among these, thrombin which is the most specific of all, is an active mitogen for chick embryo fibroblasts [182, 183], and it has been shown that only a single protein (out of about 50 which are detectable by lactoperoxidase ^{131}I-iodination techniques) on the surface of these cells is altered when they are treated with this enzyme [184]. The effect is extremely specific.

Many cell functions can be prevented by application of inhibitors of the serine proteinases, and the finding that normal calcium dependent cell functions can be stimulated by the application of calcium *in*dependent proteolytic enzymes strongly suggests that there is a common step involving limited proteolysis at the cell surface. This must occur between the attachment of the external messenger to its specific receptor and the subsequent entry of calcium into the cell: it may thus be a regulator of calcium flux.

2.5.3 Phosphatidyl inositol breakdown and resynthesis

When some tissues which have been pre-incubated in the presence of $^{32}PO_4$ are triggered into activity by application of appropriate stimuli, it is possible to detect an increased incorporation of the isotope into phosphatidyl inositol and often in phosphatidic acid also [185]. These phospholipids are present as minor components in the plasma membrane and are related in synthesis and breakdown by the metabolic cycle illustrated in Fig. 2.2.

In stimulated tissues, it is the accelerated rate of breakdown of phosphatidyl inositol to 1, 2-diacylglycerol which results in the increased rate of turnover of the PI cycle, and the detected products of the cycle are the result of resynthesis. Whilst there are no reports of phosphatidyl inositol turnover in the mast cell, there are widespread accounts of its occurence in many tissues, including all the other cell types so far mentioned in this article [185], and it is perfectly reasonable to assume that an increased rate of phosphatidyl inositol turnover is the

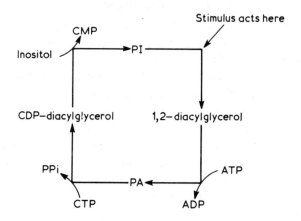

Fig. 2.2 The cycle of phosphatidyl inositol breakdown and resynthesis which is thought to occur in most tissues when exposed to specific stimuli.

accompaniment of triggering in the mast cell. The chief point to note from the present discussion is that with one exception [286], there is no evidence to link calcium (or the cyclic nucleotides) with the regulation of the PI cycle.

As for the cross linking and redistribution of surface receptors, it may be argued that the elevation in the rate of the phosphatidyl inositol cycle is a necessary and sufficient precursor step to the subsequent calcium mediated events in stimulus secretion coupling. If this is to be so then it must be shown that the elevation in the activity of the PI cycle preceeds both the expression of tissue activity (secretion, contraction, DNA synthesis etc.) and the generation of the high calcium permeability state. Certainly in some cases, elevation of the PI cycle preceeds the expression of tissue activity by a period best measured in days. For example, in lymphocytes stimulated with the lectin phytohaemagglutinin, an eighteen-fold stimulation of the PI cycle is detectable after a matter of 3 minutes [187]; the ultimate stimulation in the rate of DNA synthesis is only detectable after 48 or 72 hours. In this case, the enhancement of the PI cycle is an 'early event' indeed; but is it early enough to preceed the initial pulsed influx of calcium, which has been shown to be complete within one minute of applying the lectin? [272] (Calcium fluxes in stimulated lymphocytes are discussed in greater detail later). In certain secretory systems too, elevation of the PI cycle must be counted among the early events; for example, enhanced labelling of phosphatidic acid and phosphatidyl inositol is detectable within two seconds of stimulating platelets with thrombin [188], and in no tissue which has been critically examined does the delay amount to more than a few minutes. In antigen stimulated mast cells the calcium dependent expression of cellular activity (i.e. the secretion of histamine) is complete within one minute, and so for the PI cycle to be a controlling precursor process, it must assume its increased activity well within this time.

2.5.4 Desensitization

A fourth process which occurs in the absence of calcium is the phenomenon of desensitization [189, 190, 191]. When mast cells are treated with suitable triggering ligands *in the absence of calcium*, and then provided with calcium after a short delay, there is a time dependent decline in the amount of histamine released. With sensitized cells triggered with ovalbumin as specific antigen, the half time for this desensitization is about one minute [189], and no histamine release is elicited when the calcium is added to the antigen treated cells after a period of 5 minutes. With other triggering ligands, such as anti-rat-Ig serum, or con A, the half time for decay is extended over many minutes (Fewtrell and Gomperts, unpublished observations). Similar decay phenomena have been described for a number of other secretory tissues (*see* p 77) and it appears to be a widespread secondary response to stimulation. It may be argued that it is related to the homoeostatic maintenance of the cell and the protection of the host against the release of uncontrolled amounts of the secretory product.

2.6 REGULATION OF CALCIUM FLUX IN TRIGGERED MAST CELLS

As was stated earlier, in order to support a calcium-entry hypothesis for histamine release, one should endeavour to measure calcium fluxes arising as a consequence of stimulation by releasing ligands, and to show that secretion ensues as a consequence of introducing calcium into the cells artificially.

The measurement of calcium flux in mast cells is problematic. Although our early attempts to do this appeared to support the idea of increased

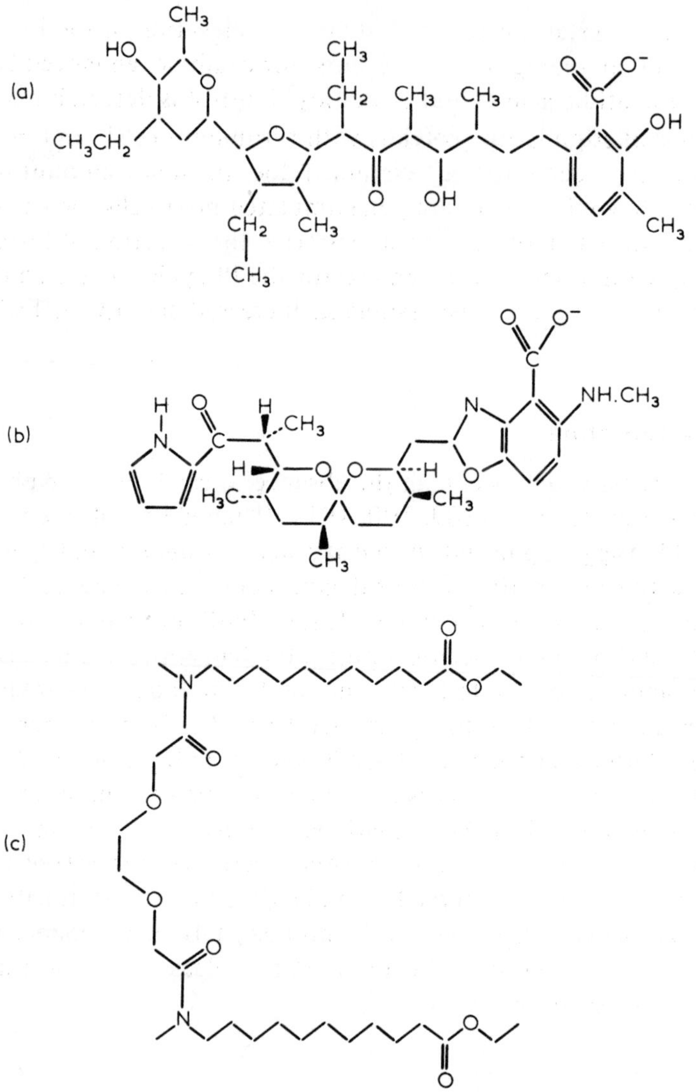

Fig. 2.3 (a) The primary structures of some carboxylate ionophores for divalent cations. (a) X537A (Hoffman la Roche); (b) A23187 (Eli Lilly); (c) a synthetic carrier which has been used to confer specificity to calcium on a membrane electrode.

calcium permeability on adding an antigen to sensitized cells [192] it is likely that much of the enhanced $^{45}Ca^{2+}$ associated with the cells was in fact bound to heparin granules, both expelled from the cells, and still bound to the cells, but accessible to the extracellular fluid. More recent

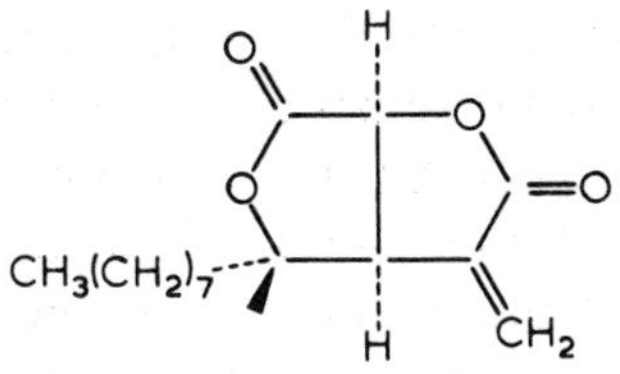

Fig. 2.3 (b) The primary structures of neutral ionophores for divalent cations: (a) beauvericin; (b) avenaciolide.

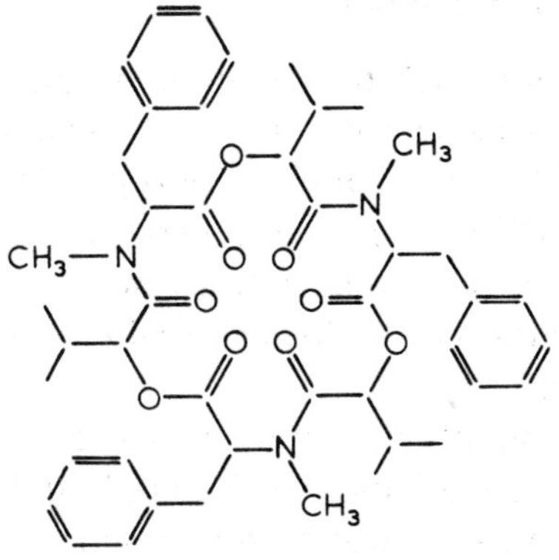

attempts [193] have shown that there is an enhanced influx of $^{45}Ca^{2+}$ when an antigen is added to metabolically inhibited (antimycin *a* treated) cells. Under these conditions, histamine is not secreted, and there are no major ultrastructural changes in the appearance of the cells, so it is unlikely that the $^{45}Ca^{2+}$ of the external fluid could come into contact with the heparin containing granules, except by passage through the cell membrane. The alternative approach, that of measuring the release of radioacivity from preloaded cells has also been exploited [40], but is open to the same criticism, and it will be necessary to repeat the experiments under inhibitory conditions, so as to separate the calcium flux from the secretory events. This remark presumes of course, that it is possible to enhance calcium permeability by application of relevant ligands to inhibited cells, but experience (see p. 91) with stimulated lymphocytes indicates that this presumtion is not without good foundation.

2.6.1 The ionophores for calcium

A much more potent approach to the whole problem of calcium
mediated stimulation became accessible with the advent of ionophores
for calcium [194, 195]. The ionophores (ϕ o ρ o ζ = bearer) are mobile
carriers of ions [196, 197]. The chemical structures of a number of
those which have been shown to be active towards calcium are illustrated
in Fig. 2.3. The ionophores of biological origin are generally products of
the genus *Streptomyces.*

The mobile carrier ionophores come in two general categories [160,
197]; the substances A23187 and X537A are both linear molecules
having a single negative charge due to the presence of a carboxyl group
(see Fig. 2.3 (a)). They cyclize by hydrogen bonding between groups at
either end of the molecule, and in so doing, they form a cavity lined
with bonding atoms having an affinity for the carried ion. In order to
carry divalent cations, two molecules of the monocarboxylate ionophores
are required (see Fig. 2.4). The A23187 calcium complex is a highly
symmetrical structure, with the calcium ion held in sixfold coordination
by two oxygens (carbonyl and carboxyl) and one nitrogen (benzoxazole)
from each ligand molecule [198, 199]. The exterior of the ionophore
calcium complex presents a lipophilic surface to the exterior environment, and it is because of this property that the ionophore is able to
dissolve the carried cation in the organic phase of the membrane
phospholipid bilayer. Solvation of cations in lipid phases is not of itself
sufficient to make a substance into an effective ionophore: many ion-
chelate complexes have this property, and they are certainly not effective
as ionophores; the problem is generally that the affinity of most well
known ligands is far too high for the carried ion ever to be released again
into an aqueous phase. The affinity of the ionophores for cations is
typically of rather a low order [197] but this of course varies with the
method used to assess it. The simplest method is by the two-phase
extraction technique of Pressman which gives values for the dissociation
constant for binding of K^+, k'_d = 90 μM for nigericin (a carboxylate
ionophore), and k'_d = 49 mM for valinomycin [197] (structurally related
to beauvericin). The affinities for divalent cations of those ionophores
which will accept them are of course much higher, but so then is the
hydration energy of the divalent cations, and the affinity for the aqueous
phase.

X537A is a carrier for divalent cations, but it is also able to transport
all the (monovalent) alkali metal cations and some primary amines,
including ethanolamine, noradrenaline and dopamine [194, 200]. Its very

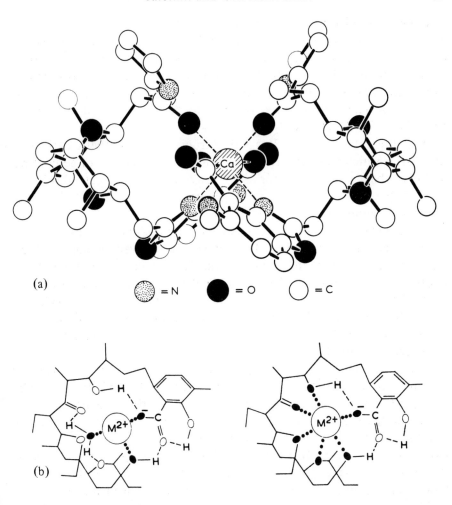

Fig. 2.4 (a) The crystal structure of Ca^{2+} (A23187)$_2$. The molecule is presented so that the pyrrole groups are at the top of the drawing, and the benzoxazole groups are at the bottom. The spiroketal pyran rings which play no direct part in metal ion bonding are at the left and right extremities. The broken lines indicates the bonds linking the central metal ion with the two ligands (from Chaney, Jones and Debono [198]).

(b) A schematic diagram illustrating the conformation of the two X537A ligands involved in divalent cation complexation. Both ligands are cyclized by head to tail hydrogen bonding but the contribution of the two of the two ligands to the structure is entirely different, resulting in an asymmetric complex. (From Pressman [194]).

low degree of selectivity may arise, at least in part, from the much greater flexibility in the chain of the molecule, compared with most other substances of this class, and this allows it to take up a variety of conformations and to accomodate ions of differing size and coordination properties. The barium salt of X537A is highly asymmetrical, with one X537A ligand offering five oxygens to the central metal ion, and the other X537A ligand only three [194, 201].

Also illustrated in Fig. 2.3(a) is the structure of a synthetic carboxylate ion carrier which was developed primarily for the purpose of generating an ion exchange membrane for use in calcium sensitive electrodes [202, 203]. It has not so far been used to generate calcium permeability in biological membranes.

The carboxylate ionophores act to dissipate ion gradients by acting as exchange diffusion carriers. They must be effectively 'loaded' as they traverse the membrane in both directions. The counter cation may be another metal cation, (e.g. Mg^{2+} would suffice for either X537A or A23187) or it may be H^+ which would combine with the carboxylate group. Because the mobile forms of the ionophore are neutral, ion transport mediated by these substances is 'blind' to such factors as membrane potential.

Beauvericin (Fig. 3(b)) is an uncharged cyclic ionophore having some resemblence to the better known valinomycin [204, 205]. It is a cyclic depsipeptide (i.e. half peptide, half ester) and has the sequence [D-α-hydroxyisovaleryl, N-methyl-L-phenylalanyl] repeated three times (see Fig. 2.3(b)). Like all the other mobile carrier ionophores, it is likely that beauvericin forms a hydrophilic cage to replace the hydration shell which surrounds the metal ion in aquesous solution. In both biological and model membrane systems (bacterial chromatophores and liposomes), beauvericin has been shown to carry calcium and the alkali metal cations [206]. In all cases, the movement of cations mediated by beauvericin dissipates the membrane potential, and it is thought that the carrier complex of beauvericin plus cation is itself a carrier of charge, as is the case for the better known K^+ ionophores valinomycin and nonactin [160]. Somewhat curiously however, the carrier complex of beauvericin plus Ca^{2+} appears to carry only a single positive charge (i.e. the same as for the monovalent cations). This is probably due to an intimate association of a single anion with the [Ca (beauvericin)$_2$]$^{2+}$ complex.

Another neutral substance which appears to be ionophoric for calcium is the antifungal dilactone, avenaciolide (from *Aspergillus avenaceus*)

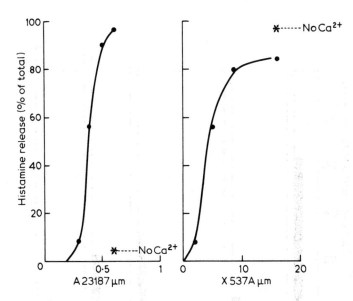

Fig. 2.5 Dose response relationships for histamine release mediated by A23187 (specific for divalent ions) and X537A (non-specific). When calcium is omitted, A23187 is no longer able to elicit release but X537A appears to be insensitive to the ionic content of the medium.

[207, 208]. This is capable of inducing calcium and magnesium permeability in rat liver mitochondria, and together with a lipid soluble anion such as thiocyanate (to maintain electroneutrality) it can render these cations soluble in organic phases [207]. The requirement for a lipophilic anion indicates that the complex between avenaciolide and divalent cations is charged, and so, like beauvericin, transport of calcium into cells mediated by this substance should be responsive to membrane potential. It is possible that questions related to the role of membrane potential in calcium activated cell phenomena could be solved with the use of ionophores of this class, but so far all the experience gained has been with the carboxylate ionophores X537A and A23187 and some of their analogues, which are electrically silent.

2.6.2 Ionophore mediated histamine release

Both X537A and A23187 are capable of eliciting histamine release from mast cells but there are important differences between the two in the way in which they achieve this effect [192]. Whilst it is difficult to be

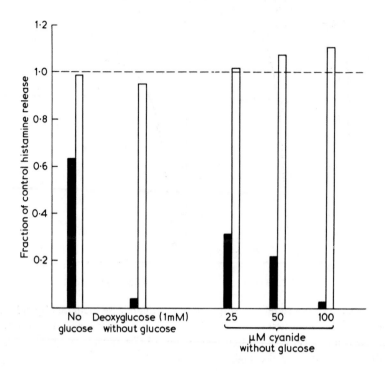

F ig. 2.6 The effects of metabolic inhibitors on ionophore mediated histamine release. The cells were preincubated with the inhibitors for 30 min before adding either A23187 (solid bars, 0.6 μM) or X537A (open bars, 17 μM). The results are expressed as a fraction of control experiments in which no inhibitors were applied. (From Foreman, Mongar and Gomperts, [192].)

precise about the effective concentration of the ionophores in the plasma membrane of the target cells, the amount of A23187 required to be added to a cell suspension in order to achieve histamine release is about an order of magnitude less than the required amount of X537A. At the higher concentration required for X537A, all the carboxylate ionophores which we have tested are capable of releasing histamine from mast cells but under these conditions, calcium ions are not required. Nigericin (a specific K$^+$ carrier [197] at 100 μM) and monensin (a sodium carrier, also capable of carrying other monovalent cations [197], at 30 μM) are both able to release histamine from mast cells, suspended in any ionic environment i.e. in the absence of the preferred ion (Fewtrell and Gomperts, unpublished observations). Release of histamine by these ionophores at rather high concentrations is not prevented by application of inhibitors of glycolysis or oxidative phosphorylation, and so it seems likely that the mechanism of release in these situations probably arises from damage to the cell membrane due to the large amount of the ionophores used. In pancreatic islets too, the action of X537A may be lytic [209], and in this case the release of the secretory product (insulin) is actually potentiated by metabolic inhibition, or by the omission of calcium.

Release of histamine mediated by A23187 is clearly dependent on the presence of calcium [192], and further resembles the ligand mediated process in being inhibited by elevated concentrations of magnesium, and by the fact that it is prevented by the application of inhibitors of glycolysis or of mitochondrial respiration (Fig. 2.6).

2.6.2 Ionophore mediated exocytosis

Examined under the electron microscope, mast cells triggered by divalent ligands, and by the calcium ionophore A23187 appear indistinguishable from each other during the early stages of release [210, 211]. The characteristic hallmark of the exocytosis process as seen in ultrathin sections, is the appearance of pentalaminar structures due to the fusion of granule membranes with the plasma membrane and with each other. These fusion structures are seen both in ligand and in A23187 triggered cells. Furthermore, we have shown that certain proteins are displaced away from the zones of fusion in both cases, so that the fused membranes are mainly composed of lipid. In particular, we found that the binding sites for anti-immunoglobulin (i.e. the cell surface IgE), and the binding sites for con A are absent from the surface of the fused membranes. When these cells are examined by freeze fracture electron microscopy [211], no intramembrane particles are found in the fracture faces of granule or plasma membranes at the points of fusion. This form of surface displacement can be distinguished from the patching phenomenon for two reasons. (1) As discussed earlier (pp 61, 62), patching (and capping) of surface receptors is not associated with movement of the particles in the fracture plane [158]: (2) The movement of the surface binding sites for IgE and for con A from fusion zones does not require cross linking by the ligand. This was demonstrated by experiments which revealed that cells triggered with A23187, and then fixed before application of ferritin-labelled con A, were unable to bind the ligand over the areas of fusion. Similarly it was possible to observe the movement of monovalent Fab-ferritin away from the areas of fusion in cells triggered with either anti-Ig or with A23187.

The movement of cell surface receptors, and the corresponding

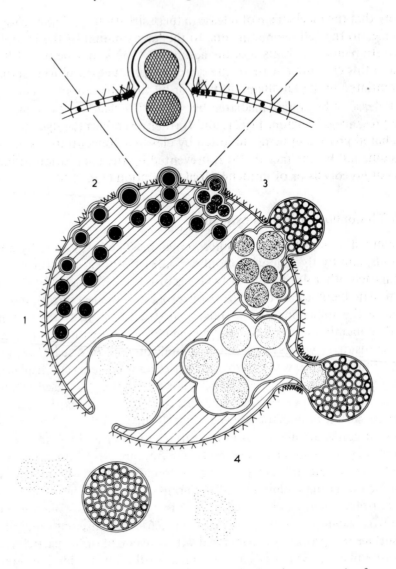

Fig. 2.7 Schematic impression of the events involved in the exocytosis of granules from rat peritoneal mast cells.

(1) The granules are initially encased tightly within the perigranular membranes and separated from the plasma membrane by the cytoplasm. The cell surface receptor sites (IgE) are displaced laterally away from the prospective fusion zones as a consequence of calcium entering the cell, and the granules approach the inner surface of the plasma membrane.

(2) The first fusion events occur at the plasma membrane (formation of

movement of the intramembrane particles in the fracture plane of the fusion zones is quite unlike the patching phenomenon: it is clearly dependent on the entry of calcium. The fused membranes then bulge out from the cell surface to form blebs ($1-2\ \mu$m diameter) which ultimately pinch-off to expose the secretory granules to the extracellular fluid. Some of the granules are actually expelled from the cell through the holes created by the shedding of the membrane blebs; the majority are retained within the periphery of the cell, but liberate their histamine by a process of cation exchange with the cations of the extracellular medium [212]. The sequence of events in exocytosis [211]: (1) lateral displacement of membrane protein away from prospective fusion sites; (2) fusion; (3) blebbing of fused membranes; (4) shedding of blebs to exteriorise granules is also of vital importance for the economy of the cell. It allows the cell to dispose of the very large amount of membrane lipid which must otherwise accrue to the surface following fusion of the granule membranes with the plasma membrane, and which would result in an uncontrolled expansion in the membrane surface area. It also allows the plasma membrane to retain at least three distinct classes of membrane protein, the surface IgE and their Fc receptors, the membrane glycoproteins, and the proteins which constitute the intercalated particles of the fracture plane. The disposal of excess material is a problem which has to be handled by all the secretory tissues, and the mechanism described for the mast cell is not the only way by which it can be overcome.

---

pentalaminar image: the central dense line is formed from the close apposition of the two cytoplasmic leaflets of the plasma membrane and the perigranular membrane). The membranes of more granules fuse with those already involved at the surface and the fusion process spreads inwards. Several granules are included within a single cavity, and the excess membrane lipid surrounding them flows to the cell surface and bulges outwards.

(3) As more granules are included within a single cavity, the excess lipid which accrues to the boundary membrane vesiculates and forms a bleb which extends out from the cell.

(4) The membrane bounding the enlarged granule cavity anneals with the plasma membrane, and the granules come into contact with the extracellular fluid, while ensuring that the cell cytoplasm is protected. The granules are either shed from the cell or retained within the exteriorized cavities from which the histamine is lost by a process of ion exchange.

In the pancreatic β-cell for example, membrane material is probably withdrawn back into the cell by the mechanism of endocytosis [213, 214].

2.6.4 ionophore mediated activity in some other systems

The ionophore A23187 has been applied to many other tissues in which it could be expected that calcium plays the role of second messenger in the triggering of activity. From many examples, one may mention the following: the secretion and aggregation of blood platelets [215, 216, 217, 218]; the secretion of insulin from the endocrine pancreas [147, 148, 209, 219]; the secretion of enzymes from exocrine pancreas [220]; the secretion of fluid and K^+ from the salivary glands [149, 221]; secretion from the thyroid [222]; secretion of dopamine from the rat striatum [200]; secretion of lysosomal enzymes from human neutrophils [223]; secretion of catecholamines from the adrenal medulla [224, 225] and of vasopressin from the neurohypophysis [226]. A23187 has also been used to regulate contraction of smooth muscle [194, 227] and single skeletal muscle fibres [228] and to increase cardiac contractility *in vitro* [229, 229a]. X537A increases cardiac contractility *in vivo* [230]. Delayed, or prolonged responses which can be triggered by application of calcium ionophores include the induction of the parthenogenic reaction in the oocytes of a large number of species [231, 232, 233, 234], and the stimulation of transformation (DNA synthesis, blastogenesis and mitosis) in the lymphocytes of pigs and humans [235, 236, 237]. The role for calcium in lymphocyte transformation is discussed later in this essay.

2.7 INHIBITION OF CALCIUM FLUX

In detail, the exocytosis of granules from mast cells triggered by application of the calcium carrier A23187 is indistinguishable from the normal process triggered by attachment of divalent ligands. However, it is misleading to overstress the similarities of function because by doing this one might be tempted to neglect the very real differences which do exist in cells released with the ionophore. The first divergence to note is in the extent of histamine released from cells triggered by ligands and by A23187. Typically, ligand stimulated cells can release as much as 40 per cent of their total histamine, but the extent of release rarely exceeds this, and it is often very much less. Release of histamine from cells treated with A23187 is of a different order, and an optimal amount of the ionophore is that amount which secures total release of histamine. The electron microscopic appearance of cells which have been totally released by incubation with A23187 and calcium is one of extensive cell damage (D. Lawson, personal communication and [160]), and this suggests that prolonged exposure to high internal concentrations of calcium is cytotoxic. The granular appearance of the cytoplasmic background normally seen in cells is absent, and there is extensive disruption and vacuolization of the Golgi apparatus. The appearance of cells released with divalent ligands is of course extensively altered by comparison with unreleased cells, but the organization of the intracellular organelles remains intact and there is no reason to suppose that the condition of the cells is anything but healthy. It is generally understood that ligand released cells are able to regenerate and after a quiescent period they may be able to undergo a second bout of secretion by exocytosis. The difference between the two modes of release is clear; ligand mediated release is self terminating, but in ionophore treated cells there is no such restraining mechanism.

2.7.1 Cell homeostasis and desensitization

Cells treated with the calcium ionophore A23187 are not subject to the phenomenon of desensitization referred to earlier [189]. Cells treated with ionophore in the absence of calcium are unable to release their histamine, but they are able to do so when the calcium level is restored (to about 1 mM) even after quite prolonged periods [189]. The ionophore by itself is not cytotoxic; it merely makes the membrane of the cell permeable to calcium. When calcium is provided, it is then enabled to enter the cell, and histamine is released normally and we may therefore conclude that desensitization of ligand treated cells is regulated by a shutting-off of the calcium permeability pathway [189]. Following attachment of the divalent ligand, the calcium pathway is opened transiently; this ensures that the cell is protected from the collossal and prolonged influx of calcium that has been shown to be toxic in the case of ionophore treated cells. Furthermore, the shutting-off of the calcium pathway permits the homeostatic functions of the cells to be reasserted, for it is only by excluding calcium from the cytosol that sensitivity to calcium can be maintained. Desensitization of tissues maintained in contact with their stimulus by a mechanism which prevents calcium entry most probably accounts for the decay phenomena observed in stimulated rat neurohypophysis [238, 239] and bovine adrenal medulla [240]. A

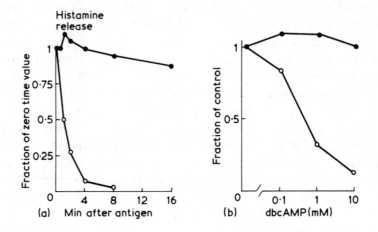

Fig. 2.8 Differential effects of inhibitory situations on histamine release from
mast cells mediated by an antigen and by A23187. (a) The course of desensitization following application of the specific antigen (egg albumin) to sensitized rat peritoneal mast cells. Addition of calcium at times after the antigen
results in a progressively reduced secretory response (o—o—o), whereas the
sensitivity of the cells to calcium after treatment with A23187 is maintained
over a considerable period of time (●—●—●). (b) Dose-effect relationship for
inhibition of histamine secretion by dibutyryl cyclic AMP. The cells were
treated with the inhibitor for 30 min before being triggered with a specific
antigen (o—o—o) or with ionophore A23187 (●—●—●), (a) from Foreman and
Garland, [189].) (b) from Foreman, Mongar, Gomperts and Garland, [244].

similar description has been applied to the 'late calcium channel' of
squid nerve which may regulate the calcium dependent transmitter
release in this tissue and other nerve terminals [241, 242].

2.7.2 The inhibitory functions of cyclic AMP

As was stated earlier, all manipulations which may be expected to elevate
the intracellular level of cyclic AMP have the effect of inhibiting ligand
mediated histamine release. This was first hinted at as early as 1936 when
Tuft and Brodsky showed that adrenaline suppresses the capacity of an
antigen administered intradermally to an allergic individual to elicit a
wheal and flare reaction in the skin [243] (though the inhibition may
have arisen, in part at least, from effects of the adrenaline on the local
capillary circulation). More directly, Schild [123] showed that adrenaline
could suppress histamine release from sensitized lung tissue *in vitro*.

In rat peritoneal mast cells, a half hour pre-incubation in the presence of dibutyryl cyclic AMP (0.5 mM) or theophylline (3 mM) will produce a 50 per cent inhibition of subsequent antigen induced histamine release, and 10 mM dibutyryl cyclic AMP reduces the release to a negligible amount. Neither theophylline nor dibutyryl cyclic AMP under these conditions are capable of reducing the extent of histamine release from mast cells treated with A23187 [244]. It is concluded that cyclic AMP acts in some way to prevent the entry of calcium into the cell by its normal route.

2.7.3 Possible modes of action of cyclic AMP

Cyclic AMP is widely understood to act as the cofactor in the activation of phosphorylase kinase in a wide range of organisms and tissues [245] and there is no reason to suppose that cyclic AMP has direct effects on the transport functions of the plasma membrane. Indeed there are a number of reasons for supposing that the effect of cyclic AMP should be indirect, and mediated via a phosphorylation catalysed by an activated form of a specific phosphorylase directed at the transport site. Certainly, cyclic AMP modulates many membrane transport systems, and in a number of cell types pretreated with ^{32}P to label the internal ATP, it has been possible to demonstrate labelling of certain membrane proteins which appear to be closely linked with transport functions. A 15 second exposure of prelabelled (^{32}P) synaptic vesicles to cyclic AMP is sufficient to label just two (out of many) of the membrane proteins which can be resolved by SDS polyacrylamide gel electrophoresis [246], and there is evidence that phosphorylation of a specific protein may be involved in the regulation of calcium transport across the membranes of the sarcoplasmic reticulum in heart muscle [247]. It is likely that this is one mechanism whereby cyclic AMP and drugs such as adrenaline which act to elevate its intracellular concentration, exert a positive inotropic effect. In this case, cyclic AMP by activating a protein kinase, raises intracellular calcium levels. The situation is complicated by a number of factors which have been discussed at length elsewhere [248, 249]. It is generally agreed that in all forms of muscle the catecholamines and cyclic AMP act both separately and synergistically to regulate the availability of calcium during the contraction-relaxation cycle. The ultimate form this takes is variable; for example, catecholamines generally cause relaxation in the longitudinal smooth muscle of the intestine, but the same substance can cause contraction of arterial smooth muscle. In the case of the arterial tissue

[250, 251], the effect seems to be a direct consequence of the effector substance, either on the membrane permeability to calcium, or on the release of calcium from intracellular stores, and cyclic AMP is probably not involved. The relaxing effect of catecholamines mediated by β-receptors on intestinal smooth muscle certainly does appear to be mediated by cyclic AMP [252] and this latter situation appears to be sufficiently similar to the situation pertaining in the mast cell to warrant further consideration in the present context. The key steps in the calcium − cyclic AMP relationship in intestinal smooth muscle have been summarized as follows [8, 249, 253, 254] : (1) adrenaline stimulates the formation of cyclic AMP by activation of adenylyl cyclase before relaxation occurs; (2) the subsequent degree of relaxation is directly proportional to the extent of cyclic AMP accumulation; (3) cyclic AMP is capable of inducing the relaxation of intestinal smooth muscle in the absence of applied adrenaline; (4) cyclic AMP stimulates the uptake of calcium by microsomal and plasma membrane fractions of broken smooth muscle preparations; (5) in intestinal smooth muscle, calcium actually inhibits cAMP phosphodiesterase, and so elevates the level of cyclic AMP; (6) a decrease in external Ca^{2+} reduces the level of cyclic AMP.

2.7.4 Cyclic AMP and desensitization

The interelationship of calcium and cyclic AMP in intestinal smooth muscle is such as to ensure that relaxation follows every bout of calcium induced contraction through exclusion of calcium from the cytoplasm. In the mast cell, it is clear that cyclic AMP is antagonistic to calcium influx from the exterior, and so one might seek for a similarity of mechanism at this point. However, there are some good reasons for believing that the control relationships between calcium and cyclic AMP are different in the mast cell, although it remains conceivable that the desensitization mechanism (≡ 'relaxation') could be put in train by a ligand or a calcium induced elevation of cyclic AMP.

When sensitized mast cells are triggered with anti-immunoglobulin there is an initial decrease (within 15 seconds) in the level of intracellular cyclic AMP which actually preceeds, and could bear a precursor relationship to the secretion of histamine [40]. The cyclic AMP then recovers and achieves its resting level within 5 min. Little is known about the purpose or the regulation of this cycle. The primary decline

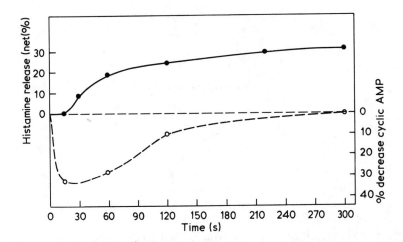

Fig. 2.9 The time course of histamine release and of changes in cyclic AMP following stimulation of mast cells with rabbit anti-rat immunoglobulin (from Kaliner and Austen, [40].)

in cyclic AMP could have a permissive effect on the entry of calcium into the cell; if this were the case, then we should want to show that it takes place in the absence of calcium. In the contrary sense, the secondary rise in cyclic AMP could possibly be the determinant of the desensitization mechanism. I do not believe this to be the case for the following reasons.

Firstly, it should be pointed out that the rate of desensitization varies with the precise nature of the stimulus applied to the cell. Thus, for cells treated with egg albumin as an antigen, $t_{1/2} \sim 1$ min [189]; but for cells treated with anti-immunoglobulin or con A, the rate of decay is much slower. We find $t_{1/2}$ for con A treated cells to be about 12 min (Fewtrell and Gomperts, unpublished observations). All these ligands stimulate the cell by attachment to the same receptors (IgE), and are understood to exert their effect through a common pathway of calcium entry. Secondly, the dose-effect (inhibition) relationship of dibutyryl cyclic AMP on all these three stimuli are similar (Fewtrell and Gomperts, unpublished observations). It is for these reasons that it seems improbable that the mechanism of desensitization is mediated by changes in the endogeneous level of cyclic AMP, and it seems more likely that the kinetic course of desensitization is controlled by the stability of the calcium pathway in the open configuration, which may be modulated differentially by individual ligands. One way of approaching this question

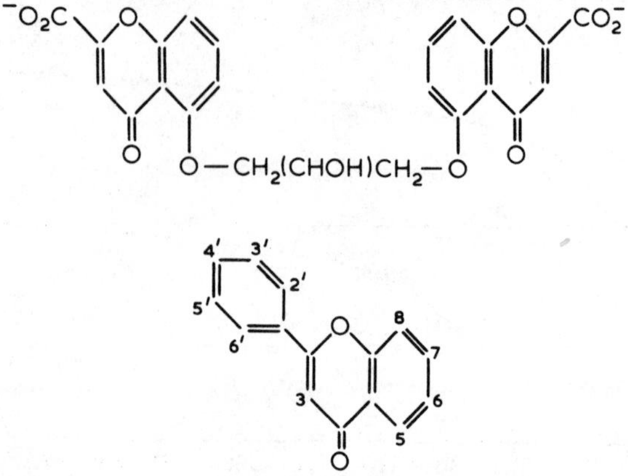

Fig. 2.10 (a) The structure of the anti-allergic drug cromoglycate.
(b) The general structure of the plant bioflavonoids. Fisetin (tetrahydroxyflavone) is substituted at positions 3,3', 4' and 7; quercetin (pentahydroxyflavone) has an additional OH group at position 5.

experimentally would be to measure the rate of decay following an initial stimulus with haptenic ligands of controlled valency (2, 3, 4 etc). Another reason for discounting a metabolic mechanism for desensitization will be discussed later, in connection with induced $^{45}Ca^{2+}$ fluxes in stimulated cells (*see* pp 90, 91).

The actual molecular mechanism whereby cyclic AMP exerts its inhibitory effect on secretion remains of central importance, and the best available clues that are presently available suggest that a search for membrane proteins which are phosphorylated as a result of elevating cyclic AMP levels could produce much desirable information.

2.7.5 The inhibitory action of cromoglycate

An important development in the treatment of asthma and the allergic diseases was the appearance of cromoglycate (the disodium salt of 1,3-bis (2-carboxycromon-5-yloxy)-2-hydroxypropane [255]). It inhibits degranulation and the secretion of histamine from mast cells [256, 257, 258] when added simultaneously with the releasing ligand. Pre-incubation of the cells with the drug is not necessary, and indeed this actually reduces the inhibitory effect so that little inhibition is demonstrable if the antigen is added after a period of 30 minutes[259, 260]. Furthermore, a second application of cromoglycate at this time is without effect; it is as if a specific inhibitor receptor on the cell surface has become 'desensitized'. A number of other pharmaceutical preparations are said to be 'cromoglycate-like' if the temporal course of inhibition is similar, and if they too are unable to reinstate the inhibitory condition of cells previously treated with cromoglycate. Cromoglycate itself has been shown to possess the properties of a phosphodiesterase inhibitor in broken cell preparations [261], and activity of this sort could of course explain its anti allergic properties. Furthermore, it was shown that cromoglycate has no inhibitory effect on cells stimulated by application of the calcium ionophore A23187 [244]. Cromoglycate, like cyclic AMP therefore appears to exert its inhibitory effect by blocking calcium entry into ligand treated cells, but the secretory apparatus of the cell remains operative, since histamine release still occurs when calcium is admitted through the back door. Some support for this statement comes from a recent report that cromoglycate also blocks the movement of $^{45}Ca^{2+}$ into stimulated mast cells [262], but unfortunately, the methods used in this investigation were probably unable to distinguish between actual stimulated flux of calcium through the mast cell membrane, and the binding of excess $^{45}Ca^{2+}$ to retained but exteriorized granules after the release event (*see* p 49). However, the drug also prevented the uptake of $^{45}Ca^{2+}$ into unstimulated mast cells, and into fibroblasts in culture, and these results can be accepted with more confidence.

As indicated above, the inhibitory action of cyclic AMP is also most probably directed at the pathway of calcium entry into the cell. There are however, sufficient points of divergence between the mode of action of cromoglycate and the recognized phosphodiesterase inhibitors (notably the immediacy, and the curious timecourse of effect) to make one wary of suggesting that they act at a common locus.

2.7.6 The inhibitory action of bioflavonoids

Certain of the plant bioflavonoids (quercetin, fisetin) also inhibit histamine release, and like cromoglycate, the effect is fully expressed when the compounds are added simultaneously with the releasing ligand (Fewtrell and Gomperts, unpublished observations); no period of pre-incubation is required, although it is not apparent that this actually reduces the inhibitory effects as is the case for cromoglycate. Unfortunately, it is quite hard to ascertain that the effect of the bioflavonoids is directed solely at the calcium entry pathway because they also have an inhibitory effect on ionophore mediated release. To be sure, a higher concentration of the inhibitor is required in order to secure a comparable degree of inhibition and under some conditions it is possible to obtain a substantial degree of inhibition of ligand triggered release, and merely a negligible effect on ionophore treated cells. The trouble centres on a direct interaction between the flavonoids and the ionophore, which effectively prevents the ionophore reaching the cell membrane. Thus we have found that the flavonoids also prevent ionophore induced fluxes of Ca^{2+} and Fe^{2+} from phospholipid liposomes loaded with these cations. That this interaction (probably a charge transfer between the benzoxazole group of A23187 and the cromone ring system of the flavonoid) takes place in the aqueous phase away from the cell surface is revealed by a red shift in the fluorescence spectrum of A23187 when it is treated, in aqueous suspension, with quercetin and certain of its homologues.

   The motivation for this investigation is two-fold. Firstly, a cursory glance at the chemical structures of cromoglycate and the plant flavones (i.e. phenyl cromones) reveals some similarities (Fig. 2.10). Clearly similarities and differences between the mode of action of cromoglycate on the one hand and the mode of action of the flavonoids on the other, could help in gaining an understanding of the mechanism of action of a drug which enjoys very widespread use [255]. This has to be seen in the context of a much wider experience of flavone biochemistry and pharmacology. Secondly, a few of the plant flavones have recently been shown to be inhibitors of the membrane ATPases. Thus they interfere with transport enzymes such as the Na^+ and K^+ ATPase of heart muscle [263], the Ca^{2+} ATPase of sarcoplasmic reticulum [264], mitochondrial (Mg^{2+}) ATPase [265] and the ATPases of chloroplasts [266] and of *Escherichia coli* [267]. At the concentrations used to prevent secretion they are not inhibitors of oxidative phosphorylation [264] and nor is quercetin an inhibitor of cyclic AMP-phosphodiesterase (Skidmore and have, personal communication). Whilst these compounds do not inhibit any of the enzymes of glycolysis individually, they do inhibit glycolytic function in certain tumour cells in which there is a defect resulting in excess ADP production due to the uncoupling of the membrane ATPases from ion transport [268, 269]. The effect of the flavonoid in this situation is to couple more closely the turnover of ATP to ion transport, so limiting availability of ADP to glycolysis; its site of action appears therefore to be in the region of the ion transport pathway itself. The experience gained in tumour cells and other systems certainly lends credence to the idea that in the flavonoids we have a class of inhibitors having certain structural characteristics in common with a widely used drug, which may be active at the very site of calcium entry into the stimulated cell. Furthermore, if this can be proved (by well controlled $^{45}Ca^{2+}$ experiments, or by the demonstration of an inhibitory effect on calcium dependent depolarization in suitable systems) it will raise the very interesting possibility that calcium fluxes, both passive into the cell, and active when coupled to the transport ATPase share a common structural pathway. We have tended to think of the calcium pathway as being created by the cross linking of surface receptors (and possibly subsequent events such as limited proteolysis, and enhanced phosphatidyl inositol turnover, (*see* pp 60–65), but now we should consider also the idea that the calcium pathway might be a permanent structural entity of the plasma membrane. In the resting cell, the calcium pathway would be coupled tightly to the calcium dependent ATPase, with the result that any calcium present in the cytosol is pumped to the exterior as soon as its concentration rises much above 10^{-8} M. As a consequence of crosslinking by a specific ligand, or by intervention further along the train of processes leading to enhanced calcium permeability, (e.g. limited proteolysis by exogenous enzymes) the calcium pathway becomes transiently decoupled from the Ca-ATPase and can be used to convey calcium ions down the concentration gradient into the cytosol. It seems possible that the flavones exert their inhibitor effect by interfering with this decoupling coupling process.

In summary, having provided strong evidence by the use of ionophore A23187, that calcium entry is the ultimate stimulus to histamine release by the mechanism of exocytosis, an analysis of the differences which exist between the ionophore and the ligand induced processes reveals that a number of inhibitory situations can be understood on the basis of a restriction of calcium movement. Inhibition by desensitization, by manipulations which can be expected to elevate the level of cyclic AMP, and by application of cromoglycate can all be explained on this basis. It is likely, but not proven, that the inhibition by the plant flavonoids can be similarly explained.

2.8 CALCIUM AND STIMULATION OF LYMPHOCYTE PROLIFERATION

All these conclusions rest on the assumption that the differences which exist between the ionophore and ligand induced mechanism of release are due to the lack of control by the cell, on ionophore mediated calcium permeability. A greater degree of confidence in these conclusions would be gained if it could be shown that the inhibitory situations are indeed matched by conditions of restricted calcium entry into the cell. However, as was discussed earlier, there are some difficulties in making such measurements, especially in a secretory cell in which the response to stimulation is as rapid and extreme as it is in the mast cell. A much better system would be one in which the initial stimulus is separated by a period of time from the ultimate expression of cellular activity. Such a system is the lymphocyte, in which attachment of triggering ligands leads to the synthesis of DNA after a delay of about 48 hours. In common with secretory cells of the immune system, DNA synthesis in lymphocytes is dependent on the presence of calcium [270], and is inhibited by manipulations which raise the level of intracellular cyclic AMP [133, 271]. Furthermore, it was reported that ionophore A23187 can be used to stimulate both pig and human lymphocytes [235, 236, 237], and in the case of pig cells it was shown that the ionophore and normal stimulating ligands synergize; that is to say, one could obtain DNA synthesis when these substances were applied together at concentrations which if applied separately, no DNA synthesis would occur [235]. It looked rather as though a similar effective stimulus was being applied to the cells by both the ionophore and by the stimulating ligand.

2.8.1 Measurement of $^{45}Ca^{2+}$ uptake

It was for reasons such as these, coupled to the very central interest which the process of lymphocyte transformation holds in the fields of immunology and cell differentiation, that we set out to describe the very early changes in lymphocyte calcium permeability following stimulation with cell specific transforming lectins [272]. The technique was simple but rather insensitive. A suspension of cells was placed above a column of silicone oil of appropriate density in a micro centrifuge tube, which was maintained at $37°C$. To the cells was added a solution of the stimulating ligand together with isotopically labelled water ($^{3}H_2O$) and calcium ($^{45}Ca^{2+}$). At the end of the incubation, the cells were sedimented by centrifugation through the column of oil. The bottom of the tube was out off, and the cell pellet dissolved out into a liquid scintillation fluid. The purpose of adding $^{3}H_2O$ to the cells was in order to measure the amount of the cells transferred through the oil layer, which proved to be very variable, probably as a result of their sticking to the walls of the plastic tube when in the presence of the transforming ligands. The results in these experiments are recorded as the ratio of cell $^{45}Ca^{2+}$ to cell $^{3}H_2O$; and this ratio was always normalized to the isotope ratio for cells treated in the absence of a stimulating ligand (adjusted to unity). The reason for using $^{3}H_2O$ as a counter for the cells was that water moves with great rapidity across membrane barriers, and even in the shortest incubations, it was likely that isotopic equilibration of ^{3}H was achieved. By substituting ^{3}H- sucrose (which is impermeant) it was possible to estimate the amount of extracellular water which was carried through the oil layer together with the cells. This proved to be considerable, and it was the far from negligible quantity of $^{45}Ca^{2+}$ always present in the entrained water space and bound to the cell surface, always present as a background, which limited the sensitivity of the method. It did however have the virtues of speed and simplicity.

2.8.2 Only T cells respond

When spleen cells (which are mainly a mixture of T and B lymphocytes) are treated with con A, there is generally an increment in the amount of $^{45}Ca^{2+}$ associated with the cells. As can be seen in Fig. 2.11, the extent of calcium uptake varies with the amount of con A applied, in a dose dependent manner, which closely resembles the dose dependence of DNA synthesis measured after a delay of 72 hours. At the high end of the dose range, there is an increment in the calcium uptake which cannot be ascribed to events relevant to cell transformation, and which could have been due to a non-specific increase in cell permeability arising from a cytotoxic concentration of the lectin.

Con A, which binds to both T and B cells, has the property of triggering cellular activity only in T cells, and so it was of the utmost interest to find out if the increment in calcium uptake by spleen cells could be ascribed to the T cells alone in this mixed-cell population. That this was indeed the case was shown by two experiments in which T cells were effectively absent. In one, the spleen cells of thymus deficient (nu/nu) mice were used. In the other, the T cells were removed by treatment of normal spleen cells with an anti-serum directed against the mouse theta (θ) alloantigen plus complement. In this way the T cells were lysed and the population could not be stimulated into DNA synthesis by con A or any of the other T cell specific mitogenic substances. Nor did these T cell deficient populations show an increase in calcium uptake when treated with T cell mitogens at relevant concentrations. Thus, the induced calcium uptake is dependent on the presence of

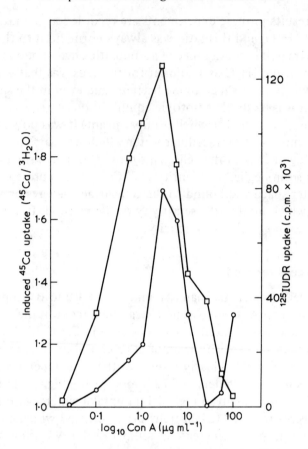

Fig. 2.11 Dose effect relationships for induced $^{45}Ca^{2+}$ uptake and DNA synthesis by con A treated spleen lymphocytes. $^{45}Ca^{2+}$ was added simultaneously with the con A, and uptake was measured during the ensuing fifteen minutes. DNA synthesis was measured by ^{131}IUDR incorporation during a twelve hour period 60 hours after the initiation of the experiment. The calcium experiment is expressed as the ratio of the con A treated $^{45}Ca^{2+}$ uptake to that of the controls, (from Freedman, Raff and Gomperts, [272].)

T cells, and the most likely explanation of this is that it is only the T cells in the mixed T + B cell population which take up calcium. In contrast, an increase in calcium uptake was never observed when the mixed cell population was treated with B specific mitogens such as bacterial lipopolysaccharide, or when T cell depleted preparations were treated with pokeweed mitogen which stimulates both T and B cells. It would

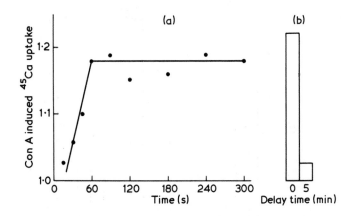

Fig. 2.12 The effect of pretreatment (30 min) with cyclic nucleotides and sodium azide on con A induced $^{45}Ca^{2+}$ uptake by mouse speen cells, (from Freedman, Raff and Gomperts, [272].)

appear that if the early increment in calcium uptake by T cells is an obligatory step in the process of proliferation, then the mechanism by which B cells are triggered is very different.

2.8.3 Inhibition of $^{45}Ca^{2+}$ uptake by cyclic AMP

The calcium uptake in con A treated cells was found to be inhibited by a number of treatments designed to elevate the level of intracellular cyclic AMP. The dose range over which dibutyryl cyclic AMP was effective in preventing calcium uptake by lymphocytes was similar to the effective dose range for the inhibition of histamine release from antigen-stimulated mast cells. Theophylline (a phosphodiesterase inhibitor) at 10^{-4} M, and cholera toxin (an activator of adenylyl cyclase) at 10^{-9} M were also inhibitory to induced calcium uptake. Dibutyryl cyclic GMP, which is known to enhance a number of lymphocyte responses, potentiated the induced uptake of calcium but none of these proceedures had any detectable effect on the flux of calcium mediated by ionophore A23187.

2.8.4 $^{45}Ca^{2+}$ uptake is transient

The con A stimulated uptake of calcium by T cells is a rapid event, which at 37°C is complete after one minute (Fig. 2.13(a)), and it was our experience of the desensitization process in mast cells which led us to

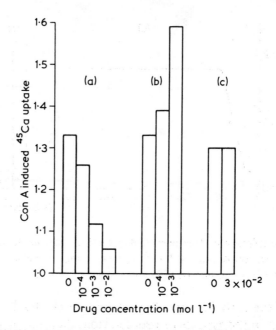

Fig. 2.13 (a) The early time course of con A induced $^{45}Ca^{2+}$ uptake by mouse spleen cells. The $^{45}Ca^{2+}$ was added together with the con A, and the cells sampled by the silicone oil centrifugation technique (as described in the text) at various times afterwards. (b) The effect of adding $^{45}Ca^{2+}$ to con A treated cells after a delay of 5 min, (from Freedman, Raff and Gomperts, [272].)

suggest that the transient nature of this process might be limited by closure of the calcium pathway. If our reasoning was correct then $^{45}Ca^{2+}$ added to con A treated cells after a delay of some minutes should not be taken up by the cells. This was found to be the case (see Fig.2.13 (b)). For reasons which have not been explained, these results differ from the results of others (working with human lymphocytes), in which the induced increment in calcium uptake was found to be undetectable before 30 min, and then prolonged [273, 274, 275].

The idea was presented earlier, that the mechanism of desensitization in mast cells, which was ascribed to a reduction in calcium permeability, is most probably not mediated by the elevation of cyclic AMP, although it does occur over an appropriate time period. In a mechanism regulated by cyclic AMP it would be expected that there must be a necessary role for ATP, and hence the apparatus of glycolysis and oxidative phosphorylation, both in order to generate the cyclic AMP and then to mediate its effects. In con A treated lymphocytes, pre-incubation with sodium azide (a mitochondrial inhibitor) at 3×10^{-2} M for 30 min produced no effect on induced calcium uptake, and it would therefore appear that the closure of the calcium pathway in stimulated cells is an event relatively independent of the metabolic process. However, it cannot be said with certainty that the regulation of calcium flux in desensitization is independent of ATP since this was not measured.

2.9 CONCLUSIONS

We now have to decide whether the 'gated' influx of calcium observed is a sufficient or necessary stimulus to the proliferative process? There are a number of reasons (some of which have already been mentioned) for believing that it may be neither sufficient nor necessary. (1) Since the removal of calcium by the addition of EGTA during the first 12 to 36 hours prevents the response, it is clear that the requirement for the presence of calcium is prolonged [276, 277]. (2) It has been shown that DNA synthesis is not prevented when calcium is added at times up to six hours after treatment with a transforming ligand [276], i.e. the calcium is there an extended requirement for calcium, but the transforming mitogen terminated. (3) The signal for transformation prolonged, for not only is there an extended requirement for calcium, but the transforming nitogen must remain in contact with its receptors on the cell surface for a period of 18 hours [278]: only after this time does it become redundant. (4) If the observed bout of enhanced calcium permeability were the sole and sufficient stimulus to DNA synthesis and proliferation, then it is surprising that this does not occur when B cells are stimulated. (5) It was noted that both pig and human lymphocytes can be stimulated by ionophore A23187. This phenomenon has not been reported for mouse lymphocytes, nor for fibroblasts grown in a monolayer culture. The proliferative stimulus to these cells shares a number of salient features of the lymphocyte system, including a dependence on the presence of calcium [279], inhibition by cyclic AMP [135, 136], and even a brief (< 1 min) period of enhanced calcium permeability following the initial stimulus to growth (Damlugi, Riley and Gomperts, unpublished observations). The reasons why ionophore A23187 stimulates some systems and not others is unclear. At the concentrations used to stimulate growth (10^{-7} M) no increment in calcium permeability was detectable by our method. The locus of activation by A23187 could be restricted to a minor component of the cell population. An alternative explanation is that pig and human lymphocytes are possibly more resistant to the cytotoxic effects of the calcium ionophore. (6) The transient increment of calcium uptake was not invariably observed (Freedman, Raff and Gomperts, unpublished observations). The calcium flux experiments were carried out during a 'season' of about four months, during which ligand induced uptake of calcium was consistently observed. After this, the response declined until it was frequently undetectable. The reasons for this have not been satisfactorily explained, but it seems likely that the variability of the response could be hormonal in origin because it was very often possible to 'rescue' the response by pretreating the cells with dibutyryl cyclic GMP. It should be added that failure to observe the response could be merely a matter of reduced magnitude rather than its total non-occurence. The method used to detect calcium uptake was relatively insensitive and always subject to a background level of radio-activity (due to the presence of calcium in the entrained water space and adsorbed at the cell surface), which was larger than the stimulated increment.

In conclusion, the measurements of calcium uptake by stimulated lymphocytes have generated a picture which closely resembles that described for triggered mast cells by a different, and indirect approach. In both systems we have seen a short-lived increment in calcium permeability following the attachment of the signalling ligand to receptors on the cell surface, and in both cases this can be prevented by manipulations designed to raise the level of intracellular cyclic AMP. In the mast cell, the presence of free internal calcium generated by this stimulated flux from the exterior is the 'second messenger' for the secretory process because, (1) there is a requirement for extracellular calcium in order for secretion to occur, (2) external calcium has to be available during the period of enhanced calcium permeability, and (3) introduction of calcium into mast cells by means of ionophore A23187 induces secretion by a mechanism which is indistinguishable from the exocytosis process triggered by divalent ligands attaching to cell surface IgE. These are the criteria which must be satisfied in order for calcium to be nominated as the sole, second messenger of cellular activation. The process of lymphocyte stimulation is surely more complex and may be controlled by a number of factors acting in sequence; it seems doubtful whether calcium entering the cell during the first minute after treatment with a signalling ligand, will prove to be one of them.

REFERENCES

1.   Ringer, S., (1883), *J. Physiol.*, **3**, 195–202.
2.   Ringer, S., (1884), *J. Physiol.*, **4**, 29–42.
3.   Locke, F.S. (1894), *Centrablat. Physiol.*, **8**, 166–167.
4.   Harvey, A.M. and MacIntosh, P.C. (1940), *J. Physiol.*, **97**, 408–416.
5.   Mann, P.J.G., Tennenbaum, M. and Quastel, J.H. (1939), *Biochem. J.* **33**, 822–835.
6.   Douglas, W.W. (1968), *Brit. J. Pharmacol. Chemother.*, **34**, 451–474.
7.   Rubin, R.P. (1970), *Pharmac. Rev.*, **22**, 389–428.
8.   Rasmussen, H., Goodman, D.B.P. and Tenenhouse, A. (1971), *Crit. Rev. Biochem.* **1**, 95–148.
9.   Bitensky, M.W. and Gorman, R.E. (1973), *Prog. Biophys.*, **26**, 409–461.
10. Parker, C.W., Sullivan, T.J. and Wedner, H.J. (1974), in *Adv. Cyc. Nuc. Res.*, Vol 4 (P. Greengard and R.A. Robison, editors), Raven Press, New York. pp. 1–79.
11. Berridge, M.J. (1975), *Adv. Cyc. Nuc. Res.*, **6**, 1–98.
12. Kamada, T. and Kinoshita, H. (1943), *Japan J. Zool.*, **10**, 469–493.
13. Heilbrunn, L.V. and Wiercinski, F.J. (1947), *J. Cell. Comp. Physiol.*, **29**, 15–28.
14. Heilbrunn, L.V. (1940), *Physiol. Zool.*, **13**, 88–94.
15. Szent György, A. (1945), *Acta, physiol. scand.*, **9**, Suppl. XXV. 1–116.
16. Szent György, A. (1946), *J. Colloid Sci.*, **1**, 1–20.
17. Miledi, R. (1973), *Proc. R. Soc. B.*, **183**, 421–425.
18. Kanno, T., Cochrane, D.E. and Douglas, W.W. (1973), *Can. J. Physiol. Pharmac.*, **51**, 1001–1004.
19. Prince, W.T. and Berridge, M.J. (1973), *J. exp. Biol.*, **58**, 367–384.
20. Pletscher, A., de Prada, M. and Tranzer, J.P. (1968), *Experientia*, **24**, 1202–1203.
21. Prince, W.T., Berridge, M.J. and Rasmussen, H. (1972), *Proc. Nat. Acad. Sci.*, **69**, 553–557.
22. Berridge, M.J., Oschman, J.L. and Wall, B.J. (1973), in *Calcium Transport in Contraction and Secretion*, (E. Carafoli, F. Clementi, W. Drabikowski and A. Margreth, editors), North Holland Publishing Co., pp. 131–138.
23. Humphrey, J.H. and Jacques, R. (1955), *J. Physiol.*, **128**, 9–27.
24. Mongar, J.L. and Schild, H.O. (1958), *J. Physiol.*, **140**, 272–284.
25. Grodsky, G.M. and Bennett, L.L. (1966), *Diabetes*, **15**, 910–913.
26. Milner, R.D.G. and Hales, C.N. (1967), *Diabetologia*, **3**, 47–49.
27. Douglas, W.W. and Rubin, R.P. (1961), *J. Physiol.*, **159**, 40–57.
28. Douglas, W.W. and Rubin, R.P. (1963), *J. Physiol.*, **167**, 288–310.
29. Woodin, A.M. and Wieneke, A.A. (1963), *Biochem. J.*, **87**, 487–495.
30. Hirsch, J.G. (1956), *J. Exp. Med.*, **103**, 589–611.
31. Becker, E.L. (1976), in *Molecular and Biological Aspects of the Acute Allergic Reactions*, 33rd Nobel Symposium, Plenum Press, London, in press.

32. Douglas, W.W., Kanno, T. and Sampson, S.R. (1967), *J. Physiol.*, **191**, 107–121.
33. Dean, P.M. and Matthews, E.K. (1970), *J. Physiol.*, **210**, 253–264.
34. Dean, P.M. and Matthews, E.K. (1970), *J. Physiol.*, **210**, 265–275.
35. Douglas, W.W. and Poisner, A.M. (1962), *J. Physiol.*, **162**, 385–392.
36. von Hattingberg, H., Kuschinsky, G. and Rahn, K.H. (1966), *Naunym-Schmiedbergs Arch. Exp. Path. Pharmak.*, **253**, 438–443.
37. Case, R.M. and Clausen, T. (1973), *J. Physiol.*, **235**, 75–102.
38. Malaisse, W.J. and Brisson, G.R. (1972), *Diabetologia*, **8**, 56 (abstract).
39. Malaisse, W.J. (1973), *Diabetologia*, **9**, 167–173.
40. Kaliner, M. and Austen, K.F. (1974), *J. Immunol.*, **112**, 664–674.
41. Blaschko, H., Comline, R.S., Schneider, F.H., Silver, M. and Smith, A.D. (1967), *Nature*, (Lond.), **215**, 58–59.
42. Lagunoff, D. (1972), *Biochem. Pharmac.*, **21**, 1889–1896.
43. Lettvin, J.Y., Pickard, W.F., McCulloch, W.S. and Pitts, W. (1964), *Nature*, (Lond.), **202**, 1338–1339.
44. van Breemen, C. and McNaughton, E. (1970), *Biochem. Biophys. Res. Comm.*, **39**, 567–574.
45. van Breemen, C., Farinas, B.R., Casteels, R., Gerba, P., Wuytack, F. and Deth, R. (1973), *Phil. Trans. R. Soc. Ser. B.*, **265**, 57–71.
46. Jobsis, F.F. and O'Connor, M.J. (1966), *Biochem. Biophys. Res. Commun.*, **25**, 246–252.
47. Ashley, C.C. and Caldwell, P.C., (1974), *Biochem. Soc. Symp.*, **39**, 29–50.
48. Baker, P.F. (1972), *Prog. Biophys.*, **24**, 177–223.
49. Caswell, A.H. (1972), *J. Membrane Biol.*, **7**, 354–364.
50. Täljedal, I. (1974), *Biochim. Biophys. Acta*, **372**, 154–161.
51. Engbaek, L. (1952), *Pharmac. Rev.*, **4**, 396–414.
52. Jenkinson, D.H. (1957), *J. Physiol.*, **138**, 434–444.
53. Dodge, F.A. and Rahaminoff, R. (1967), *J. Physiol.*, **193**, 418–432.
54. Burton, R.F. and Loudon, J.R. (1972), *J. Physiol.*, **220**, 363–381.
55. Foreman, J.C. and Mongar, J.L. (1972), *J. Physiol.*, **224**, 753–769.
56. Stossel, T.P. (1973), *J. Cell Biol.*, **58**, 346–356.
57. Golstein, P. and Gomperts, B.D. (1975), *J. Immunol.*, **114**, 1264–1268.
58. Williams, R.J.P. (1970), *Quart. Rev. Chem. Soc.*, **24**, 331–365.
59. Dixon, M. and Webb, E.C. (1964), *Enzymes*, Second edition, Longmans, London, pp. 421 *et seq.*
60. Schatzmann, H.J. (1973), *J. Physiol.*, **235**, 551–569.
61. MacLennan, D.H. (1970), *J. Biol. Chem.*, **245**, 4508–4518.
62. Ebashi, S. (1974), *Essays in Biochemistry* **10**, 1–36, Academic Press, London.
63. Wessels, N.K., Spooner, B.S. and Luduena, M.A. (1973), in *Locomotion of Tissue Cells* (Ciba Foundation Symposium No. 14. New Series), Elsevier/North Holland, pp. 53–76.
64. Pollard, T.D. and Weihing, R.R. (1974), in *CRC Critical Reviews in Biochemistry*, Vol. 2, CRC Press Inc., Cleveland, Ohio, pp. 1–65.

65. Burridge, K. and Phillips, J.H. (1975), *Nature*, **254**, 526–529.
66. Röhlich, P. (1975), *Exp. Cell. Res.*, **93**, 293–298.
67. Stossel, T.P. and Pollard, T.D. (1973), *J. Biol. Chem.*, **248**, 8288–8294.
68. Lazarides, E. (1975), *J. Cell. Biol.*, **65**, 549–562.
69. Tanaka, H. and Hatano, S. (1972), *Biochim. Biophys. Acta*, **257**, 445–451.
70. Cohen, I. and Cohen, C. (1972), *J. Molec. Biol.*, **68**, 383–387.
71. Fine, R.E., Blitz, A.L., Hitchcock, S.E. and Kaminer, B. (1973), *Nature, New Biol.*, **245**, 182–186.
73. Shibata, N., Tatsumi, N., Tanaka, K., Okamura, Y. and Senda, N. (1972), *Biochim. Biophys. Acta*, **256**, 565–576.
74. Nachmias, V. and Asch, A. (1974), *Biochem. Biophys. Res. Commun.*, **60**, 656–664.
75. Weisenberg, R.C. (1972), *Science*, **177**, 1104–1105.
76. Olmsted, J.B. and Borisy, G.G. (1973), *Ann. Rev. Biochem.*, **42**, 507–540.
77. Borisy, A.A., Olmsted, J.B., Marcum, J.M. and Allen, C. (1974), *Fedn. Proc.*, **33**, 167–174.
78. Orr, T.S.C., Hall, D.E. and Allison, A.C. (1972), *Nature*, (Lond.), **236**, 350–351.
79. Allison, A.C. (1973), in *Cell Locomotion* R. Porter and D.W. Fitzsimmons, editors, Ciba Symposium 13 (New Series) Elsevier/North Holland, pp. 109–148.
80. Dean, P.M. (1975), *J. Theoret. Biol.*, **54**, 289–308.
81. Allison, A.C., Davies, P. and de Petris, S. (1971), *Nature, New Biol.*, **232**, 153–155.
82. Carter, S.B. (1967), *Nature*, (Lond.), **213**,, 261–264.
83. Spudich, J.A. and Lin, S. (1972), *Proc. Nat. Acad. Sci.*, **69**, 442–446.
84. Puszkin, C., Puszkin, S., Lo, W. and Tanenbaum, S.W. (1973), *J. Biol. Chem.*, **248**, 7754–7761.
85. Forer, A.I., Emmerson, I. and Behnke, O. (1972), *Science*, **175**, 774–776.
86. Lazarides, E. and Weber, K. (1974), *Proc. Nat. Acad. Sci.*, **71**, 2268–2272.
87. Wessells, N.K., Spooner, B.S., Ash, J.F., Bradley, M.O., Luduena, M.A., Taylor, E.L., Wrenn, J.T. and Yamada, K.M. (1971), *Science* **171**, 135–143.
88. Schroeder, T.E. (1970), *Z. Zellforsch. Mikrosk. Anat.*, **109**, 431–449.
89. Spooner, B.S. and Wessells, N.K. (1970), *Proc. Nat. Acad. Sci.*, **66**, 360–364.
90. Pesanti, E. and Axline, S.G. (1975), *J. Exp. Med.*, 1030–1040.
91. Williams, J.A. and Wolff, J. (1971), *Biochem. Biophys. Res. Commun.*, **44**, 422–425.
92. Zigmond, S.H. and Hirsch, J.G. (1972), *Science*, **176**, 1432–1434.
93. Zigmond, S.H. and Hirsch, J.G. (1972), *Expl. Cell Res.*, **73**, 383–393.
94. Kletzien, R.F., Perdue, J.F. and Springer, A. (1972), *J. Biol. Chem.*, **247**, 2964–2966.
95. Sanger, J.W. and Holtzer, H. (1972), *Proc. Nat. Acad. Sci.*, **69**, 253–257.
96. Mizel, S.B. and Wilson, L. (1972), *J. Biol. Chem.*, **247**, 4102–4105.

97. McDaniel, M.L., King, S., Anderson, S., Fink, J. and Lacy, P.E. (1974), *Diabetologia,* **10**, 303–308.
98. Everhart, L.P. and Rubin, R.W. (1974), *J. Cell Biol.,* **60**, 434–441.
99. van Obberghen, E., Somers, A., Devis, G., Ravazzola, M., Malaisse-Lagae, F., Orci, L. and Malaisse, W.J. (1975), *Diabetes,* **24**, 892–901.
100. Davis, A.T., Estensen, R. and Quie, P.A. (1971), *Proc. Soc. Exp. Biol. Med.,* **137**, 161–164.
101. Zurier R.B., Hoffstein, S. and Weissmann, G. (1973), *J. Cell Biol.,* **58**, 27–41.
102. Zurier, R.B., Hoffstein, S. and Weissmann, G. (1973), *Proc. Nat. Acad. Sci.,* **70**, 844–848.
103. Zurier, R.B., Weissmann, G., Hoffstein, S., Kammerman, S. and Tai, H.H. (1974), *J. Clin. Invest.,* **53**, 297–309.
104. Bauduin, H., Stock, C., Vincent, D. and Grenier, J.F. (1975), *J. Cell Biol.,* **66**, 165–181.
105. Palade, G. (1975), *Science,* **189**, 347–358.
106. Douglas, W.W., Poisner, A.M. and Rubin, R.P. (1965), *J. Physiol.,* **179**, 130–137.
107. Schneider, F.H., Smith, A.D. and Winkler, H. (1967), *Brit. J. Pharmac. Chemother.,* **31**, 94–104.
108. Gordon, J.L. (1975), in *Lysosomes in Biology and Pathology,* Vol. 4, (J.T. Dingle and R.T. Dean, editors) North Holland, American Elsevier, pp. 3–31.
109. Holmsen, H. (1975), in *Biochemistry and Pharmacology of Platelets,* (G.V.R. Born, editor), Ciba Foundation Symposium 35 (New Series), Elsevier/North Holland, pp. 175–196.
110. Lagunoff, D., Pritzl, P. and Mueller, L. (1970), *Exp. Cell Res.,* **61**, 129–132.
111. Palade, G.E. (1959), in *Subcellular Particles,* (T. Hayashi, editor), Ronald, New York, pp. 64–83.
112. Ishizaka, K. and Ishizaka, T. (1967), *J. Immunol.,* **99**, 1187–1198.
113. Orange, R.P., Stechschulte, D.J. and Austen, K.F. (1970), *J. Immunol.,* **105**, 1087–1095.
114. Bach, M.K., Bloch, K.J. and Austen, K.F. (1971), *J. Exp. Med.,* **133**, 752–771.
115. Mota, I. (1957), *Brit. J. Pharmacol., Chemother.,* **12**, 453–456.
116. Humphrey, J.H., Austen, K.F. and Rapp, H.J. (1963), *Immunology,* **6**, 226–245.
117. Keller, R. (1973), *Clin. Exp. Immunol.,* **13**, 139–147.
118. Levine, B.B. (1965), *J. Immunol.,* **94**, 111–131.
119. Ishizaka, K. and Ishizaka, T. (1968), *J. Immunol.,* **103**, 588–595.
120. Magro, A.M. and Alexander, A. (1974), *J. Immunol.,* **112**, 1757–1761.
121. Siraganian, R.P., Hook, W.A. and Levine, B.B. (1975), *Immunochemistry,* **12**, 149–157.
122. Lawson, D., Fewtrell, C., Gomperts, B. and Raff, M.C. (1975), *J. Exp. Med.,* **142**, 391–402.

123. Schild, H.O. (1936), *Quart. J. Exp. Physiol.*, **26**, 166–179.
124. Lichtenstein, L.M. and Margolis, S. (1968), *Science*, **161**, 902–903.
125. Assem, E.S.K. and Schild, H.O. (1969), *Nature*, (Lond.), **224**, 1028–1029.
126. Koopman, W.J., Orange, R.P. and Austen, K.F. (1970), *J. Immunol.*, **105**, 1096–1102.
127. Bourne, H.R., Lichtenstein, L.M. and Melmon, K.L. (1972), *J. Immunol.*, **108**, 695–705.
128. Johnson, A.R., Moran, N.C. and Mayer, S.E. (1974), *J. Immunol.*, **112**, 511–519.
129. Kaliner, M. and Austen, K.F. (1974), *Biochem. Pharmac.*, 763–771.
130. Cox, J.P. and Karnovsky, M.L. (1973), *J. Cell Biol.*, **59**, 480–490.
131. Ignarro, L.J. (1975), in *Lysomes in Biology and Pathology*, Vol. 4, (J.T. Dingle and R.T. Dean, editors), Elsevier/North Holland, pp. 481–523.
132. Zurier, R.B., Tynanat, N. and Weissman, G. (1973), *Fedn. Proc.*, **32**, 2982, (abstract).
133. Abell, C.W. and Monahan, T.M. (1973), *J. Cell Biol.*, **59**, 549–550.
134. Haslam, R.J. (1975), in *Biochemistry and Pharmacology of Platelets*, (G.V.R. Born, editor), Ciba Foundation Symposium (New Series), Elsevier/North Holland, pp. 121–143.
135. Johnson, G.S. and Pastan, I. (1972), *J. Natn. Cancer Inst.*, **48**, 1377–1383.
136. Otten, J., Johnson, G.S. and Pastan, I. (1972), *J. Biol. Chem.*, **247**, 7082–7087.
137. Barger, G. and Dale, H.H. (1910), *J. Physiol.*, **41**, 19–59.
138. Jaanus, S.D. and Rubin, R.P. (1974), *J. Physiol.*, **237**, 465–476.
139. Guidotti, A. and Costa, E. (1973), *Science*, **179**, 902–904.
140. Brisson, G.R., Malaisse-Lagae, F. and Malaisse, W.J. (1972), *J. Clin. Invest.*, **51**, 232–241.
141. Carchman, R.A., Jaanus, S.D. and Rubin, R.P. (1971), *Molec. Pharmacol.*, **7**, 491–499.
142. Birmingham, M.K., Elliott, F.H. and Valero, H.L.P. (1953), *Endocrinology*, **53**, 687–689.
143. Birmingham, M.K., Kurlents, E., Lane, R., Muhlstock, B. and Traikov, H., (1960), *Can. J. Biochem. Physiol.*, **38**, 1077–1085.
144. Peron, F.A. and Koritz, S.B. (1958), *J. Biol. Chem.*, **233**, 251–259.
145. Jaanus, S.D., Carchman, R.A. and Rubin, R.P. (1972), *Endocrinology*, **91**, 887–893.
146. Haksar, A., Maudsley, D.V., Peron, F.A. and Bedigian, E. (1976), *J. Cell Biol.*, **68**, 142–153.
147. Charles, M.A., Lawecki, J., Pictet, R. and Grodsky, G.M. (1975), *J. Biol. Chem.*, **250**, 6134–6140.
148. Karl, R.C., Zawalich, W.S., Ferrendelli, J.A. and Matschinsky, F.M. (1975), *J. Biol. Chem.*, **250**, 4575–4579.
149. Prince, W.T., Rasmussen, H. and Berridge, M.J. (1973), *Biochim. Biophys. Acta*, **329**, 98–107.

150. Taylor, R.B., Duffus, W.P.H., Raff, M.C. and de Petris, S. (1971), *Nature, New Biol.,* **233**, 225–229.
151. de Petris, S. and Raff, M.C. (1973), *Nature, New Biol.,* **241**, 257–259.
152. Becker, K.E., Ishizaka, T., Metzger, H., Ishizaka, K. and Grimley, P.M. (1973), *J. Exp. Med.,* **138**, 394–409.
153. Wallach, D.F.H. and Zahler, P.H. (1966), *Proc. Nat. Acad. Sci.,* **56**, 1552–1559.
154. Allison, A.C. (1974), in *Control of Proliferation in Animal Cells,* (B. Clarkson and R. Baserga (eds.),), Cold Spring Harbor Laboratory, pp. 447–459.
154a. Gingell, D. (1976), in *Mammalian Cell Membranes,* Vol. 1, (G.A. Jamieson and D.M. Robinson (eds.),), Butterworths, London, pp. 198–223.
155. Mueller, P. (1975) in Energy Transducing Mechanisms, (E. Racker editor), MTP International Review of Science, Biochemistry, Series 1 Vol 3. Butterworths, London, pp. 75–120.
156. Pinto da Silva, P. and Nicolson, G.L. (1974), *Biochim. Biophys. Acta,* **363**, 311–319.
157. Ji, T.H. and Nicolson, G.L. (1974), *Proc. Nat. Acad. Sci.,* **71**, 2212–2216.
158. Karnovsky, M.J. and Unanue, E.R. (1973), *Fedn. Proc.,* **32**, 55–59.
159. Bretscher, M. and Raff, M.C. (1975), *Nature,* (Lond.), **258**, 43–49.
160. Gomperts, B.D. (1976), *The Plasma Membrane: Models for Structure and Function,* Academic Press, in press.
161. Becker, E.L. and Austen, K.F. (1964), *J. Exp. Med.,* **120**, 491–506.
162. Becker, E.L. and Austen, K.F. (1966), *J. Exp. Med.,* **124**, 379–395.
163. Mahler, H.R. and Cordes, E.H. (1966), *Biological Chemistry,* Harper International Edition, p. 290.
164. Kaliner, M. and Austen, K.F. (1973), *J. Exp. Med.,* **138**, 1077–1094.
165. Diamant, B. (1976), in *Molecular and Biological Aspects of the Acute Allergic Reactions,* 33rd. Nobel Symposium, Plenum Press, London, in press.
166. Uvnas, B. and Antonsson, J. (1963), *Biochem. Pharmacol.,* **12**, 867–873.
167. Saeki, K. (1964), *Japan J. Pharmacol.,* **14**, 375–390.
168. Sasaki, J. (1975), *Japan J. Pharmacol.,* **25**, 311–324.
169. Lagunoff, D., Chi, E.Y. and Wan, H. (1975), *Biochem. Pharmacol.,* **24**, 1573–1578.
170. Luscher, E.F. and Massini, P. (1975), in *Biochemistry and Pharmacology of Platelets,* (G.V.R. Born, editor), Ciba Foundation Symposium 35, (New Series), Elsevier, Exerpta Medica, North Holland, pp. 5–15.
171. Detwiler, T.C., Martin, B.C. and Feinman, R.D. (1975), in *Biochemistry and Pharmacology of Platelets,* (G.V.R. Born, editor), Ciba Foundation Symposium 35 (New Series), Elsevier, Exerpta Medica, North Holland, pp. 77–91.

172. Pearlman, D.S., Ward, P.A. and Becker, E.L. (1969), *J. Exp. Med.*, **130**, 745–764.
173. Ward, P.A. and Becker, E.L. (1967), *J. Exp. Med.*, **125**, 1001–1020.
174. Ward, P.A. and Becker, E.L. (1967), *J. Exp. Med.*, **125**, 1021–1030.
175. Saito, M., Yoshizawa, T., Aoyagi, T. and Nagai, Y. (1973), *Biochem. Biophys. Res. Commun.*, **52**, 569–575.
176. Moreau, P., Dornand, J. and Kaplan, J.G. (1976), *Canad. J. Biochem.*, **53**, 1337–1341.
177. Burger, M.M. (1970), *Nature*, **227**, 170–171.
178. Vischer, T.L. (1974), *J. Immunol.*, **113**, 58–62.
179. Kaplan, J.G. and Bona, C. (1974), *Exp. Cell Res.*, **88**, 388–394.
180. Noonan, K.D. and Burger, M.M. (1973), *Exp. Cell Res.*, **80**, 405–414.
181. Noonan, K.D. (1976), *Nature*, (Lond.), **580**, 573–576.
182. Chen, L.B. and Buchanan, J.M. (1975), *Proc. Nat. Acad. Sci.*, **72**, 131–135.
183. Teng, N.N.H. and Chen, L.B. (1975), *Proc. Nat. Acad. Sci.*, **72**, 413–417.
184. Teng, N.N.H. and Chen, L.B. (1976), *Nature*, (Lond.), **259**, 580–582.
185. Michell, R.H. (1975), *Biochim. Biophys. Acta*, **415**, 81–147.
186. Lennon, A.M. and Steinberg, H.R. (1973), *J. Neurochem.*, **20**, 337–345.
187. Fisher, D.B. and Mueller, G.C. (1971), *Biochim. Biophys., Acta*, **248**, 434–448.
188. Lloyd, J.V., Nishizawa, E.E. and Mustard, J.F. (1973), *Brit. J. Haematol*, **25**, 77–79.
189. Foreman, J.C. and Garland, L.G. (1974), *J. Physiol.*, 381–391.
190. Baxter, J.H. and Adamik, R. (1975), *J. Immunol.*, **114**, 1034–1041.
191. Diamant, B., Grosman, N., Stahlskov, P. and Thomle, S., (1974), *Int. Archs. Allergy, appl. Immun.*, **47**, 412–424.
192. Foreman, J.C., Mongar, J.L. and Gomperts, B.D. (1973), *Nature*, (Lond.), **245**, 249–251.
193. Foreman, J.C., Hallett, M.B., and Mongar, J.L. (1975), *Brit. J. Pharmac.*, **55**, 283P–284P.
194. Pressman, B.C. (1973), *Fedn. Proc.*, **32**, 1698–1703.
195. Reed, P.W. and Lardy, H.A. (1972), *J. Biol. Chem.*, **247**, 6970–6973.
196. Pressman, B.C., Harris, E.J., Jagger, W.S. and Johnson, J.H. (1967), *Proc. Nat. Acad. Sci.*, **58**, 1949–1956.
197. Pressman, B.C. (1968), *Fedn. Proc.*, **27**, 1283–1288.
198. Chaney, M.O., Jones, N.D. and Debono, M. (1976), *J. Antibiotics*, **29**, (in press).
199. Deber, C.M. and Pfeiffer, D.R. (1976), *Biochemistry*, **15**, 132–140.
200. Holz, R.W. (1975), *Biochim. Biophys. Acta*, **375**, 138–152.
201. Johnson, S.M., Herrin, J., Liu, S.J., and Paul, I.C. (1970), *J. Amer. Chem. Soc.*, **92**, 4428–4435.
202. Simon, W., Morf, W.E. and Meier, P.C. (1973), *Structure and Bonding*, **16**, 113–160.
203. Amman, D., Pretsch, E. and Simon, W. (1973), *Helv. Chim. Acta*, **56**, 1780–1787.

204. Roeske, R.W., Issac, S., King, T.E. and Steinrauf, L.K. (1974), *Biochem. Biophys. Res. Commun.*, **57**, 554–561.
205. Hamilton, J.A., Steinrauf, L.K. and Braden, B. (1975), *Biochem. Biophys. Res. Commun.*, **64**, 151–156.
206. Prince, R.C., Crofts, A.R. and Steinrauf, L.K. (1974), *Biochem. Biophys. Res. Commun.*, **59**, 697–703.
207. Harris, E.J. and Wimhurst, J.M. (1973), *Nature, New Biol.*, **245**, 271–273.
208. Brookes, D., Tidd, B.K. and Turner, W.B. (1963), *J. Chem. Soc.*, 5385–5391.
209. Hellman, B. (1975), *Biochim. Biophys. Acta*, **399**, 157–169.
210. Kagayama, M. and Douglas, W.W. *J. Cell Biol.*, **62**, 519–526.
211. Lawson, D., Gilula, N.B., Fewtrell, C.M.S., Gomperts, B.D. and Raff, M.C. (1976), in *Molecular and Biological Aspects of the Acute Allergic Reactions*, 33rd Nobel Symposium, Plenum Press, London, (in Press).
212. Uvnäs, B. (1974), *Fedn. Proc.*, **33**, 2172–2176.
213. Orci, L., Malaisse-Lagae, F., Ravazzola, M., Amherdt, M. and Renold, A.E. (1973), *Science*, **181**, 561–562.
214. Orci, L. (1974), *Diabetologia*, **10**, 163–187.
215. Feinman, R.D. and Detwiler, T.C. (1974), *Nature*, (Lond.), **249**, 172–173.
216. Massini, P. and Luscher, E.F. (1974), *Biochim. Biophys. Acta*, **372**, 109–121.
217. White, J.G., Rao, G.H.R. and Gerrard, J.M. (1974), *Amer. J. Pathol.*, **77**, 135–149.
218. Gerrard, J.M., White, J.G. and Rao, G.H.R. (1974), *Amer. J. Pathol.*, **77**, 151–166.
219. Wollheim, C.B., Blondel, B., Truehart, P.A., Renold, A.E. and Sharp, G.W.G. (1975), *J. Biol. Chem.*, **250**, 1354–1360.
220. Eimerl, S., Savion, N., Heichal, O. and Selinger, Z. (1974), *J. Biol. Chem.*, **249**, 3991–3993.
221. Selinger, Z., Eimerl, S. and Schramm, M. (1974), *Proc. Nat. Acad. Sci.*, **71**, 128–131.
222. Grenier, G., van Sande, J., Glick, D. and Dumont, J.E. (1974), *FEBS Letters*, **49**, 96–99.
223. Smith, R.J. and Ignarro, L.J. (1975), *Proc. Nat. Acad. Sci.*, **72**, 108–112.
224. Garcia, A.G., Kirpekar, S.M. and Prat, J.C. (1975), *J. Physiol.*, **244**, 253–262.
225. Cochrane, D.E., Douglas, W.W., Mouri, T. and Nakazato, Y. (1975), *J. Physiol.*, **252**, 363–378.
226. Nordmann, J.J. and Currell, P.A. (1975), *Nature*, (Lond.), **253**, 646–647.
227. Murray, J.J., Reed, P.W. and Fay, F.S. (1975), *Proc. Nat. Acad. Sci.*, **72**, 4459–4463.
228. Hainaut, K. and Desmedt, J.E. (1974), *Nature*, (Lond.), **252**, 407–408.
229. Schaffer, S.W., Safer, B., Scarpa, A. and Willaimson, J.R. (1974), *Biochem. Pharmacol.*, **23**, 1609–1617.
229a. Holland, D.R., Steinberg, M.J. and Armstrong, W. McD. (1975), *Proc. Soc. Exp. Biol. Med.*, **148**, 1141–1145.

230. de Guzman, N.T. and Pressman, B.C. (1974), *Circulation*, **49**, 1072–1077.
231. Steinhardt, R.A. and Epel, D. (1974), *Proc. Nat. Acad. Sci.*, **71**, 1915–1919.
232. Chambers, E.L., Pressman, B.C. and Rose, B. (1974), *Biochem. Biophys. Res. Commun.*, **60**, 126–132.
233. Steinhardt, R.A., Epel, D., Carroll, E.J. and Yanagimachi, R. (1974), *Nature*, (Lond.), **252**, 41–43.
234. Scheutz, A.W. (1975), *J. Exp. Zool.*, **191**, 433–440.
235. Maino, V.C., Green, N.M. and Crumpton, M.J. (1974), *Nature*, **231**, 324–327.
236. Luckasen, J.R., White, J.G. and Kersey, J.H. (1974), *Proc. Nat. Acad. Sci.*, **71**, 5088–5090.
237. Hovi, T., Allison, A.C. and Williams, S.C. (1976), *Exp. Cell Res.*, **97**, 92–100.
238. Nordmann, J.J. (1975), *J. Physiol.*, **249**, 38P–39P.
239. Nordmann, J.J. (1975), in *Calcium Transport in Contraction and Secretion*, (E. Carafoli, F. Clementi, W. Drabikowski and A. Margreth, editors). North Holland/American Elsevier, pp. 281–286.
240. Rink, T.J. and Baker, P.F. (1975), in *Calcium Transport in Contraction and Secretion*, (E. Carafoli, F. Clementi, W. Drabikowski and A. Margreth, editors), North Holland/American Elsevier, pp. 235–242.
241. Baker, P.F., Meves, H. and Ridgway, E.B. (1973), *J. Physiol.*, **231**, 527–548.
242. Baker, P.F. and Glitsch, H.G. (1975), *Phil. Trans. Roy. Soc. Ser. B.*, **270**, 389–409.
243. Tuft, L. and Brodsky, M.L. (1936), J. *Allergy*, **7**, 238–249.
244. Foreman, J.C., Mongar, J.L., Gomperts, B.D. and Garland, L.G. (1975), *Biochem. Pharmac.*, **24**, 538–540.
245. Kuo, J.F. and Greengard, P. (1969), *Proc. Nat. Acad. Sci.*, **64**, 1349–1355.
246. Greengard, P. (1976), *Nature*, (Lond.), **260**, 101–108.
247. Ueda, T., Maeno, H. and Greengard, P. (1973), *J. Biol. Chem.*, **248**, 8295–8305.
248. Bulbring, E. and Kuriyama, H. (1973), *Phil. Trans. Roy. Soc. Ser. B.*, **265**, 113–121.
249. Anderson, R., Lundholm, L., Mohme-Lundholm, E. and Nilsson, K. (1972), in *Adv. Cyc. Nuc. Res.* Vol. 1. (P. Greengard and G. Robison editors), Raven Press, New York.
250. Hudgins, P.M. (1969), *J. Pharmac. Exp. Ther.*, **170**, 303–310.
251. Namm, D.H. (1971), *J. Pharmac. Exp. Ther.*, **178**, 299–310.
252. Levine, R.A., Cafferata, E.P. and McNally, E.F. (1967), *Rec. Adv. Gastroenterology*, **1**, 408.
253. Andersson, R. and Mohme-Lundholm, E. (1969), *Acta Physiol. Scand.*, **77**, 372–384.
254. Andersson, R. and Mohme-Lundholm, E. (1970), *Acta Physiol. Scand.*, **79**, 244–261.
255. Pepys, J. (1969), in *Hypersensitivity Diseases of the Lungs due to Fungi and Organic Dusts*, S. Karger, Basel.

256. Morse, H.C., Austen, K.F. and Bloch, K.J. (1969), *J. Immunol.*, **102**, 327–337.
257. Assem, E.S.K. and Mongar, J.L. (1970), *Int. Archs. Allergy, appl. Immun.*, **38**, 68–77.
258. Orr, T.S.C., Pollard, M.C., Gwilliam, J. and Cox, J.S.C. (1971), *Life Sciences*, **10**, 805–812.
259. Kusner, E.J., Dubnick, B. and Herzig, D.G. (1973), *J. Pharmac. Exp. Ther.*, **184**, 41–46.
260. Evans, D.T., Marshall, P.W. and Thomson, D.S. (1975), *Int. Arch. Allergy, appl. Immun.*, **49**, 417–427.
261. Roy, A.C. and Warren, B.T. (1974), *Biochem. Pharmac.*, **23**, 917–920.
262. Spataro, A.C. and Bosmann, H.B. (1976), *Biochem. Pharmacol.*, **25**, 505–510.
263. Carpenedo, F., Bortignon, C., Bruni, A. and Santi, R. (1969), *Biochem. Pharmac.*, **18**, 1495–1500.
264. Suolinna, E-M., Buchsbaum, R.N. and Racker, E. (1975), *Cancer Res.*, **35**, 1865–1872.
265. Lang, D.R. and Racker, E. (1974), *Biochim. Biophys. Acta*, **333**, 180–186.
266. Deters, D.W., Racker, E., Nelson, N. and Nelson, H. (1975), *J. Biol. Chem.*, **250**, 1041–1047.
267. Futai, M., Sternweis, P. and Heppel, L. (1974), *Proc. Nat. Acad. Sci.*, **7**, 2725–2729.
268. Suolinna, E-M., Lang, D. and Racker, E. (1974), *J. Nat. Cancer Inst.*, **53**, 1515–1519.
269. Racker, E. (1975), in *Energy Transducing Systems*, (E. Racker editor), MTP International Review of Science, Biochemistry Series 1 Vol. 3. Butterworths, pp. 163–183.
270. Alford, R.H. (1970), *J. Immunol.*, **104**, 698–703.
271. Smith, J.W., Steiner, A.L. and Parker, C.W. (1971), *J. Clin. Invest.*, **50**, 442–448.
272. Freedman, M.H., Raff, M.C. and Gomperts, B.D. (1975), *Nature*, (Lond.), **255**, 378–382.
273. Allwood, G., Asherson, G.L., Davey, M.J. and Goodford, P.J. (1971), *Immunology*, **21**, 509–516.
274. Whitney, R.B. and Sutherland, R.M. (1972), *Cell Immunol.*, **5**, 137–147.
275. Whitney, R.B. and Sutherland, R.M. (1973), *J. Cell Physiol.*, **82**, 9–19.
276. Diamantstein, T. and Ulmer, A. (1975), *Immunology.*, **28**, 121–125.
277. Whitney, R.B. and Sutherland, R.M. (1972), *J. Cell Physiol.*, **80**, 329–337.
278. Novogrodsky, A. and Katchalski, E. (1971), *Biochim. Biophys. Acta*, **228**, 579–583.
279. Balk, S.D. (1971), *Proc. Nat. Acad. Sci.*, **68**, 271–275.
280. Kuo, I.C.Y. and Coffee, C.J. (1976), *J. Biol. Chem.*, **251**, 1603–1609.

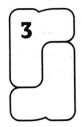

3 Cell Traffic

MARIA de SOUSA
Sloan-Kettering Institute for Cancer Research,
1275 York Avenue,
New York 10021

Acknowledgements

I wish to acknowledge my colleagues Drs D.M.V. Parrott, A. Freitas and Marlene Rose for much lively discussion on the subject of 'why cells travel', Mr Callender for the preparation of the illustrations and Miss G. Marshall for the typescript. E. Maya's patience and knowledge of the alphabet took care of the list of references. Work by the author was supported by C.R.C. Grant SP 1832.

Receptors and Recognition, Series A, Volume 2
Edited by P. Cuatrecasas and M.F. Greaves
Published in 1976 by Chapman and Hall, 11 New Fetter Lane, London EC4P 4EE
© Chapman and Hall

INTRODUCTION

Standing on a bridge watching a stream flow, one can easily understand the general traffic and ultimate destination of Autumn leaves and floating ducks. A child can. Anyone can. The wonder is how a cell, a world so much smaller than a leaf or a duck, can achieve such delicate balance between inanimate and intelligent behaviour, enter a flow inert yet leave it, defying the ruling laws of traffic of inert things. But some cells do; it is their job to, as it is ours to wonder.

3.1 GENERAL BEHAVIOUR OF CELLS IN FLOWS

Most basic data on the behaviour of cells in a flow have been obtained with red cells in a blood flow, examined either in artificial tubular circuits (Merrill *et al.*, 1965; Fahraeus and Lindquist, 1931) or in *in vivo* circuits running in structures that can be transilluminated easily, such as the cheek pouch in the hamster (Fulton *et al.*, 1946; Greenblatt *et al.*, 1969), the mesentery (Zweifach and Intaglietta, 1968; Bloch, 1962), the bat's wing (Wiedeman, 1962, 1963), the heart chambers in the cat, the dog, and the turtle (Tillich *et al.*, Hellberg *et al.*, 1972; Tillmanns *et al.*, 1974).

From these studies it has been demonstrated that the velocity of flow varies in an inverse relationship to the cross-sectional area of the blood vessels. Thus, for example, in man the velocity of flow is slowest in the capillary vessel region (0.07 cm s^{-1}) which has a cross-sectional area of approximately 5000 cm^2 in contrast with the much smaller cross-sectional area of the largest arteries and veins, $20-30$ cm^2, where the velocity of blood flow is as high as 50 cm s^{-1} in the arteries and 40 cm s^{-1} in the veins (Fig. 3.1).

At present it is possible to measure not only the velocity of the flow and of the cells in the flow, but also the velocities of cells in different positions across the same vessel (reviewed by Nims and Irwin, 1973). In a streamline flow, cells in the midline position move fastest; in Table 3.1

105

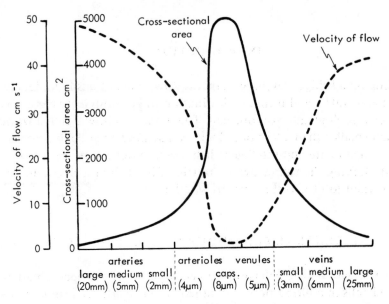

Fig. 3.1 Diagrammatic representation of the relationship between cross-sectional area and blood flow at different levels of the systematic blood circuit. Modified from Berne, R.M. and Levey, M.N. (1972) *Cardiovascular Physiology* (2nd edn.) C.V. Mosby, St. Louis, and Navaratnam, V. (1975) *'The human heart and circulation',* Academic Press, New York.

Table 3.1* Cell velocity in 60 μm hamster vessel

Distance from centre (μm)	Velocity (cm s^{-1})
0.0	0.36
10.0	0.29
20.0	0.20
30–29	0.12

*   From Charm and Kurland, 1974.

an example of the relationship between velocity and distance from the centre in a 60 μm vessel of the hamster cheek pouch is presented (Berman and Fuhnor, 1966; quoted in Charm and Kurland, 1974). This type of information is most valuable because it has made possible the determination of the values of shear rate at the vessel walls (Table 3.2). This in turn has given support to the idea that blood flow does not occur in even streamlines with a uniform cell concentrations, but that a

Table 3.2* Estimates of wall shear rates† in various vessels in man

Vessel	Average velocity (cm s^{-1})	Diameter (cm)	γw (s^{-1})
Aorta	40	2.5	155
Artery	45	0.4	900
Arteriole	5	0.005	8000
Capillary	0.1	0.0008	1000
Venule	0.2	0.002	800
Vein	10	0.5	160
Vena cava	38	3.0	100

* From Charm and Kurland, 1974.
† (γw = 8 ($\bar{V}/D$)).

layer of decreased red cell concentration is present near the wall (Charm and Kurland, 1974).

On the other hand, studies of the behaviour of white cells in the blood flow of the hamster cheek pouch and mouse mesentery (Fulton *et al.*, 1946; Florey, 1958; Atherton and Born, 1972) have indicated that white cells, unlike red cells, roll along the vessel wall in the small vessels; the mean velocity of rolling granulocytes is directly proportional to the mean blood flow velocity within a 300–1000 μm s^{-1} range (Atherton and Born, 1973). As an infection develops, the white cells tend to stick to the wall and form a pavement in the vessel, reducing its cross-section considerably. Increased rolling granulocyte counts can also be induced by a local application of chemotactic agents (Atherton and Born, 1972). The interest of this kind of study resides in the proximity to physio-pathological reality, the kind of reality that cannot be derived from purely *in vitro* models of cell flow.

One example of the discrepancy between results obtained *in vivo* and *in vitro* is the difference in resistance to blood flow in arteries and glass tubes of similar physical characteristics, i.e. diameter and length (Whitaker and Winton, 1933; Kurland *et al.*, 1968). Resistance to blood flow *in vivo* is constantly less than resistance to blood flow *in vitro*. The question of electric charge has been investigated as a possible explanation for the lower resistance *in vivo*, and indeed negatively charged tubing affords lower flow resistance to red cells than positive or neutral tubing (Charm and Kurland, 1974). Electric charge, however, acts at small distances (10 Å) and it is unlikely that the lower resistance of negatively charged tubing is due exclusively to repulsion between cell and tube wall.

In a study of the shear stress required for detachment of human erythrocytes from artificial surfaces, Hochmutt *et al.,* (1972) found that a minimum critical value of 10 dyn cm^{-2} was needed. This value, however, could be reduced to 5 or 2.5 if the surfaces were coated with albumin or fibrinogen.

In conclusion, studies of the flow of cells in artificial circuits, or in those limited vascular circuits that can be transilluminated are most valuable in giving us a general picture of the behaviour of cells in traffic. For example, the well documented phenomenon of the axial migration of cells, and existence of a plasma layer between flowing cells and vessel wall, is in theory identical to the axial migration exhibited by rubber discs in glycerol of the same density (Muller, 1941). However, as Charm and Kurland (1974) point out: 'In the microcirculation, cells do not necessarily move in a straight path but frequently bounce from the wall suggesting that he so-called plasma layer is in fact statistical'. This suggests in addition, that the flow of a single cell may in its appearance mimic the simplicity of the flow of a rubber disc, but the complexity involved in the interaction between that cell, other cells in the flow, and other cells in the endothelium lining the vascular circuit, must be very different from the interaction of a rubber disc with another rubber disc on a glass wall.

3.2 THE FLEXIBLE GEOMETRY OF THE BLOOD AND LYMPH CIRCUITS

The only place where measurements such as 8 μm for a capillary and 50 μm for a post-capillary venule are going to stay geometrically unchanged is in the graph itself (Fig. 3.1). In the reality of physiological kinetics the variation in the diameter of a single venule from one moment to the next can be considerable. Thus Tillmanns *et al.,* (1974), in a study of the microcirculation of the ventricle of the dog and the turtle, found that the inside diameters of arterioles in dogs were 15–29 μm during systole and 20–36 μm during diastole, and the diameters of venules were 11–20 μm during systole and 15–29 μm during diastole. In the turtle the differences were still more marked; arteriole diameters were in the range 12–26 μm during systole and 14–40 μm during diastole, and venules were 9–15 μm during systole and 13–21 μm during diastole. Red cell velocity too varied in the arterioles and venules during systole and diastole (Fig. 3.2). In the capillaries and venules peak velocities were observed during systole, and in the arterioles during diastole.

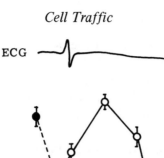

ECG

Fig. 3.2 Phasic red cell velocities in arterioles (●– – – –●) capillaries
(△—·—·△) and venules (○——○) of the turtle ventricle, during systole and
diastole. Red cell velocity in capillaries and venules increased during whole
systole from 909 ± 35 to 1988 ± 130, and from 1516 ± 77 to 3358 ± 113,
respectively. Velocity in the arterioles decreased from 3317 ± 258 to 1734
± 155. ECG represented. Modified from Tillmanns *et al.,* (1974).

The decrease in red cell velocity observed in the capillaries during
early diastole is related to the transient increase in capillary cross-
sectional area occurring at this time. The relationship between red cell
velocity (V) and capillary cross-sectional area (A) can be expressed by
applying the equation of continuity:

$$A_1 V_1 \; = \; A_2 V_2 \tag{1}$$

$$V_2 \; = \; A_1 V_1 / A_2 \tag{2}$$

in which A_1 and V_1 represent the cross-sectional area of arterioles and
red cell velocity in the arterioles during diastole, and A_2 and V_2 represent
the same parameters for the capillaries. Capillary red cell velocity (V_2)
thus decreases as cross-sectional area (A_2) increases.
  Evidence indicating that the geometry of vessels is flexible is not

derived exclusively from studies of the microcirculation in transilluminat-able structures. Herman *et al.,* (1972), in an elegant study of the blood microcirculation in the rabbit popliteal lymph node during a primary immune response, compared the microangiograms of thick sections (250–350 μm) prepared at different times (3 h to 75 days) following the injection of an antigen in the footpad with the micro-angiograms of the contralateral node draining the control uninjected footpad.

The microcirculation of the resting lymph node consists of a system of arteries with an average diameter of 20–40 μm branching into a capillary network most prominent under the subcapsular sinus and in the medullary cords, made up of capillaries with an average diameter of 11 μm, which in turn get together (in groups of 2 to 5) to form the post-capillary venules, with an average diameter of 15–30 μm. Post-capillary venules are most numerous in those areas of the cortex with a rich capillary supply, i.e. the subcapsular region. The post-capillary venules join to form large venules which enter one of the main veins situated in the cortico-medullary junction with a diameter of 140 μm.

Slight increases in the diameter of the capillaries in the submarginal sinus region were first observed 6 h after antigen stimulation. At 24 h both the capillary diameter and the density of distribution of the capil-laries had increased; by 48 h an increasingly dense capillary network was observed virtually without avascular areas in the draining node. The medullary cord capillaries had at this time an average diameter of 15–20 μm. By the 5th day many capillaries in the medullary cords had a diameter of 25 μm, an increase of over 100% compared with the aver-age diameter of 11 μm seen in the resting state. In addition to the finding of an increased density distribution of capillaries and increased diameters, Herman *et al.,* also noticed extravasation of the contrast medium they used in a considerably higher number of immunized than control nodes. They thought the extravasation might indicate increased permeability or decreased pressure resistance of the small blood vessels during the im-mune response, a finding of considerable significance to reported changes in lymphocyte traffic occurring after antigen administration (see Section 3.4.3.2).

Having illustrated how flexible the geometry of the microcirculation is, especially at the points with largest cross-sectional areas, slowest flow, and most important for the interaction of the circulating cells with endothelia, I shall summarize briefly the gross anatomy of the three major circuits utilized by cells in traffic: the systemic, the pulmonary, and the lymphatic circuits.

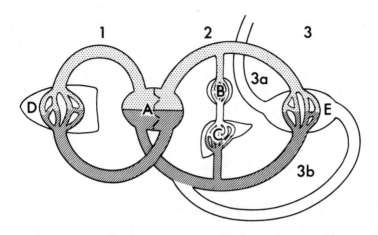

Fig. 3.3 Diagram illustrating the pulmonary (1) and systemic (2) blood circuits, the peripheral (3a) and central (3b) lymph circuits utilized by red and white cells in the mammal. Oxygenated blood (▨) leaves the heart (A) into the systemic circuit, which includes the spleen (B), the liver (C) and the total mass of the peripheral lymphoid tissue (E). From the periphery the deoxygenated blood (▨) returns to the heart via the lungs (D). The peripheral lymphoid tissues in addition to receiving lymphocytes from the blood (see Fig. 3.13) also act as a lymph filter via the afferent lymphatics (3a), the efferent lymphatics eventually drain into major lymphatic ducts (3b).

In the mammals, where the majority of the experiments on cell traffic *in vivo* have been undertaken, the blood leaves the heart left ventricle oxygenated, goes into the systemic circuit, and returns to the right atrium and right ventricle to enter the pulmonary circuit de-oxygenated. After oxygenation in this circuit, the blood re-enters the left atrium, the left ventricle, and the systemic circuit (Fig. 3.3).

The lymphatic circuit consists of a meshwork of terminal lymph capillaries which run parallel to the regional capillaries and venules, leading into larger lymph vessels which coalesce to form four major lymph trunks: the jugular, the subclavian, the mediastinal, and the lumbar. All the lymphatics of the lower part of the body and the left upper half drain into a large duct known as the thoracic duct, and the lymphatics from the right upper half regions drain into the right lymph duct; both major lymph ducts connect with the large thoracic veins. Lymph nodes are capsulated aggregates of lymphoid and phagocytic cells distributed along the major lymphatic routes. Since the time of the Greeks, lymph nodes have been visualized as fixed structures in the lymphatic circuit acting as filters of the humors (Hippocrates; see Littré, 1853) or of pathogenic bacteria (see Florey, 1958). In the light of our present knowledge of lymphocyte traffic (see Section 3.4) a lymph node is a traffic jam strategically placed between the blood and the lymph circuits (Fig. 3.3). Before reaching the jam points, a cell injected intravenously passes at least three major capillary networks: the pulmonary, the splenic, and the hepatic networks.

3.3 RED AND WHITE CELL TRAFFIC

3.3.1 Historical development of separate roots of interest

In the subject of cell traffic, as in any subject, the questions to which we are at present addressing ourselves are largely the reflection of the way in which the subject itself has developed, and of the specialists that through history became interested in it. Thus, it is not surprising to find that most of the work on red cell traffic is to be found in haematology and clinical journals, relating mostly to the question of the mechanism of anaemia and distribution of labelled red cells in the systemic circulation in man (Jandl *et al.*, 1957; Mollison, 1962; Brown, 1973; Rifkind, 1966). Work on white cell traffic, on the other hand, can be divided sharply into questions related to the traffic of granulocytes and the traffic of lymphocytes; the former developed closely linked to the study of inflammation and the discovery of chemotaxis, and is therefore mostly concerned with factors influencing cell traffic in sites of inflammation (see review by Grant, 1973). The origin in modern times of the work on lymphocyte traffic (Gowans, 1957, 1959; Gowans and Knight, 1964) preceded closely the break-through in understanding of the role of different classes of lymphocytes in the immune response (see Section 3.4). Most of the work is to be found in immunology journals, reflecting the main preoccupations of immunologists in their own field: specificity of the immune response and mechanism of cell recruitment from the circulation (see review by Ford, 1975).

Whereas in the field of red cell traffic, anaemia has been the dominant factor governing the analysis of factors influencing red cell distribution, in the field of lymphocyte traffic lymphopenia is hardly ever mentioned; however, as discussed later (Section 3.6.1) the underlying factors controlling traffic of all cell types may be very similar, and our present divisions of interest and understanding are the price paid for the enjoyment of specialized knowledge and restricted interest.

3.3.2 Factors influencing red cell traffic

A red cell can be visualized as a flexible structure consisting of two major components: the intracellular haemoglobin fluid and a flexible membrane that separates the cell contents from the external medium. A normal red cell adapts easily to flow by virtue of its natural deformability; deformability is a function of the ratio between the viscosity of the external medium and the viscosity of the haemoglobin itself. Red cell traffic thus must be influenced by factors that reduce cell deformability and increase blood viscosity. In addition, factors that alter the rigidity of the red cell membrane and modify the cell membrane's interaction with other cells should also affect the progress of a red cell in the blood flow.

3.3.1.1 Increased haemoglobin viscosity: sickle cell disease

Sickled cells have a deformability which is much lower than normal as the result both of altered membrane flexibility and of higher viscosity of the reduced S–S haemoglobin compared with the viscosity of the reduced haemoglobin of normal red cells (Dintenfass, 1964). It seems that, even in full oxygenation, capillary flow of sickled cells is impaired (Chien *et al.*, 1970). This is reflected in the frequent finding of thrombosis, vascular occlusions, and infarction in a number of capillary beds (Uszoy, 1964; Finch, 1972), the most important of which is the renal bed because of the severe consequences of impaired kidney function (Schlitt and Keitel, 1960).

3.3.1.2 Spherocytosis

Spherocytosis is the morphological expression of a damaged red cell, which is seen in naturally occurring hereditary spherocytosis and in a number of other blood diseases, especially in association with auto-immune haemolytic anaemia. Spheroid forms can be induced *in vitro* by heat damage (Ham *et al.*, 1948; Kimber and Lander, 1964) and by exposure of red cells to antibody or complement (C_3) in the presence of macrophages (Jandl *et al.*, 1957; Lay and Nussenzweig, 1968; Huber *et al.*, 1969). In the naturally occurring disease, spherocytes have a diminished surface area and volume, sometimes with a diameter as small as 3 μm and only 20% of the original cell volume.

Heat damage

Studies in man of the fate of ^{51}Cr-labelled red cells damaged by heat have shown that the cells are removed rapidly from the circulation mainly into the spleen and liver (Crome and Mollison, 1964; Kimber

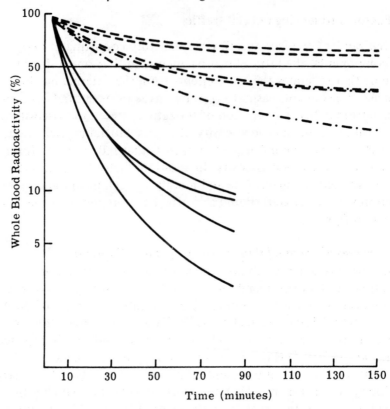

Fig. 3.4 Rate of clearance from blood of ^{51}Cr-labelled heat damaged red cells in control (—) and splenectomized (- - -, —·—·) patients. Modified from Marsh *et al.*, (1966).

and Lander, 1964; Marsh *et al.,* 1966). In a study analysing the effect of heating time on distortion of red cell morphology, clearance time, and organ distribution, Kimber and Lander (1964) demonstrated that heating at 50°C for 10 min caused mainly spherocytic change and slow removal of the cells by the spleen; heating at the same temperature for 20 min resulted in the formation of spheroid and fragmented forms distributed both in the spleen and liver, and heating for much larger periods of time (1 and 2 h) produced agglutination of red cells which were removed preferentially by the liver. In this study Kimber and Lander (1964) also found that after moderate heating, e.g. 50°C for 20 min, the clearance rate could be expressed as two exponential curves, a rapid component with a half-clearance time of 4–8 min and a slow component with a clearance time of 60–100 min, presumably indicating a differential susceptibility of red cell populations to heat damage.

In a study comparing the fate of autologous [51] Cr-labelled red cells in control and splenectomized patients, Marsh *et al.*, (1966) found that red cell clearance in the absence of the spleen was much prolonged (Fig. 3.4). The slow decrease in blood radioactivity was associated with some uptake of the damaged cells by the liver. The exact mechanism of clearance of spherocytes from the blood by the spleen is not fully understood.

Analysing the question of spherocytosis from a rheological point of view, Erslev and Atwater (1963) reported the existence of an exponential relation between mean corpuscular haemoglobin and whole blood viscosity; for example, blood viscosity increases sharply when the mean corpuscular haemoglobin reaches values similar to those found in naturally occurring spherocytosis (34%). The high blood viscosity could influence red cell transit in the spleen capillaries, creating ideal physical conditions of interaction between the circulating cell and surrounding macrophages in the red pulp cords, which would lead to its removal from the circulation.

Modification of red cell surface by antibody
The pioneer observation by Jandl *et al.*, (1957) that anti-D coated red cells adhered to leucocytes *in vitro*, acquiring a spherocytic form, confirmed a decade later by work on the effect of IgG and C3 on red cell macrophage binding (Lay and Nussenzweig, 1968), opened up a new area of understanding of the molecular basis of red cell sequestration (Brown and Lachman, 1971; Brown *et al.*, 1970; Schreiber and Frank, 1972a, Atkinson *et al.*, 1973).

Two hours after the administration of [51] Cr-unsensitized autologous red cells in the guinea-pig (Atkinson *et al.*, 1973) the highest concentrations of labelled cells is found in the circulating blood (87%), with small percentages in the liver (6.3%) and spleen (3.5%). Pretreatment of the erythrocytes with IgM (117 Cl-fixing sites per cell) results in clearance of 50–75% of the injected cells by the liver within 10 min. The presence of the sequestered cells in the liver is transient, and one-third of the sequestered cells eventually return to the circulation. Temporary sequestration of [51] Cr-labelled red cells in the liver has also been described in the rabbit following injection of labelled C3-coated cells (Brown and Lachman, 1971); in these experiments, re-entry of the labelled cells in circulation coincided with the decay of functionally active C3 *in vivo*

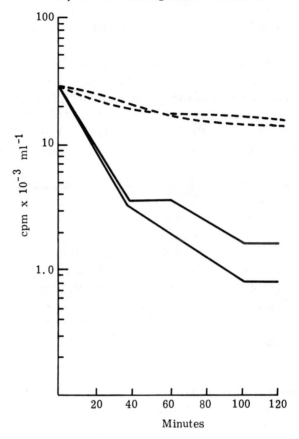

Fig. 3.5 Rate of clearance from the blood of ^{51}Cr-labelled guinea-pig red
cells sensitized with 17 IgG Cl-fixing sites per cell. Modified from Atkinson
et al., (1973).

and with the appearance of small spherocytes in the peripheral blood.
Pretreatment of guinea-pig erythrocytes with IgG (Atkinson *et al.*, 1973)
leads to a slower erythrocyte sequestration occurring mostly in the
spleen (70%) and less markedly in the liver (25%).

Atkinson *et al.*, (1973) studied also the effect of splenectomy on the
clearance of IgG pretreated ^{51}Cr-labelled cells. The results at an IgG
concentration of 17 IgG Cl-fixing sites (Fig. 3.5) are strikingly similar
to those found by Marsh *et al.*, (1966) in splenectomized patients given
radiolabelled heat-damaged cells (Fig. 3.4). Increasing the number of
complement fixing sites in the red cells to 34 and to 90, however,
completely abolished the differences between the behaviour of the intact

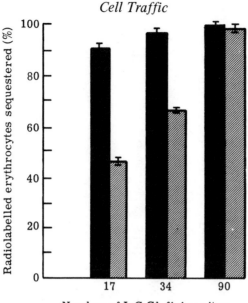

Fig. 3.6 Influence of numbers of IgG Cl-fixing sites per cell in degree of sequestration (clearance from blood) of ^{51}Cr-labelled erythrocytes in normal (■) and splenectomized (▨) guinea-pigs. Clearance increased as level of sensitization was increased. Modified from Atkinson *et al.*, (1973).

and the splenectomized recipients (Fig. 3.6). Splenectomy did not alter the fate of IgM coated cells. These results indicate that sequestration of red cells depends on the class of antibody present on the circulating cell surface and is not determined by the spleen being the organ specifically responsible for red cell clearance. Clearance of the sensitized red cells in splenectomized hosts was done mostly by the liver; there was virtually no participation of the lungs in the process of sequestration (Table 3.3).

The finding that red cells modified by antibody are not sequestered in the lungs is an interesting one. The pulmonary capillary network is the first network encountered by cells injected intravenously, and traffic of other cells through the lungs is sometimes transiently delayed (Freitas and de Sousa, 1975, 1976a). Perhaps red cell traffic delay through the lungs would have such serious consequences for the host that interaction between the alveolar capillary and the red cell evolved to allow rapid transit. Dr. Freitas, in my laboratory, analysed this question by tracing the fate of ^{51}Cr-labelled red cells after treatment *in vitro* with substances known to cause transient lung delay of lymphocyte traffic. The results (included in Table 3.3) indicate that none of the

Table 3.3 Effect of various *in vitro* treatments of red cells on their *in vivo* fate following intravenous injection in experimental animals

Reference	Species	Treatment	Time after injection (min)	Distribution change			
				Blood	Lungs	Liver	Spleen
Brown *et al.*, 1970	Rabbit	$C_{43}(5)$	8–15	ND	→	→	→
Atkinson *et al.*, 1973	Guinea-pig	3 IgG₁ C₁ fixing sites	5–120	↓	ND	→	→
		17 IgG₁ C₁ fixing sites	5–120	↓	ND	→	→
		34 IgG₁ C₁ fixing sites	5–120	↓	↔	→	→
		117 IgM fixing sites	5–120	↓	ND	→	↔
Freitas, 1976	Mouse	Trypsin	60	↓	↔	→	→
		PL-A	60	↓	↔	→	↔
		Con-A	60	↓	↔	↔	→

↔ : No change; → : increase; ↓ : decrease; ND : not done.

treatments with trypsin, Concanvalin A (Con-A), PL-A, or neuraminidase, alter red cell traffic through the lung.

3.3.1.3 Modification of the host

There have been few experiments designed to investigate how modification of the host environment influences red cell distribution, with the exception of the work on the effect of splenectomy. Atkinson *et al.*, (1973), in an attempt to define the experimental basis of the known beneficial effect of steroids in the management of auto-immune haemolytic anaemia, investigated the effect of cortisone treatment on sequestration of ^{51} Cr-labelled red cells sensitized *in vitro* with IgG or IgM antibody, in guinea-pigs. Cortisone treatment was effective in reducing clearance of red cells sensitized with IgG at low levels (3, 7, and 17 Cl-fixing sites) when given 4 days before the labelled red cell transfer. Cortisone pretreatment had no effect on the clearance of red cells at high levels of sensitization with IgG (34 Cl-fixing sites/cell).

The exact relevance of these results to human auto-immune haemolytic anaemia remained unresolved from this study, and the way in which the cortisone influenced red cell sequestration is also not clearly understood, mainly because the action of corticosteroids has so many facets that it is difficult to know with precision which one is responsible for the reduced clearance by the liver in the case of IgM sensitization and by the spleen in the case of IgG antibody.

In summary, red cell traffic is influenced by physical factors, namely blood viscosity and cell deformability; if traffic is slowed down in sinusoid networks with rich macrophage linings, such as the splenic red pulp or the liver, alteration of the red cell membrane by antibody or complement further delays the transit of the cell through the capillary network, often leading to its destruction by phagocytosis. In reviewing a subject it is perhaps just as important to stress the work that has been done as to wonder about the work that could have been done if historically the subject had evolved not from interest in anaemia or blood rheology but from interest in membrane components responsible for cell recognition. One looks forward to the day when more work on the effect of enzymes, surfactants, plant lectins, and viruses on red cell traffic will lead to an integration with what is already known about the effect of these substances on the traffic of other cell types, particularly lymphocytes (see Table 3.6).

3.3.3 Factors influencing granulocyte traffic

Considering that 50—70% of the leucocytes in the blood circulation are granulocytes, and that these cells always appear first at a site of inflammation (Florey, 1970) or shortly after the administration of an antigen (Herman *et al.*, 1972), and even after application of agents which elicit delayed hypersensitivity reactions (de Sousa and Parrott, 1969), the amount of work that has been done on granulocyte traffic is surprisingly small (see Grant, 1973). The bulk of the work in this field seems to have been concentrated on factors influencing granulocyte locomotion rather than traffic.

Although admittedly one cannot make a long journey between two distant points by a known route without the effort of getting up from the departure point and the relief of sitting down at arrival, the scope of this review is more about the nature of the journey and the things that make it sufficiently eventful to be remembered than about getting up or sitting down. Reviews of current issues in the field of leucocyte adhesion, chemotaxis, and locomotion have been published recently (Wilkinson, 1976).

One seems to know little about the regular journey of granulocytes. They are in the bone marrow, and in the mouse there are additional specific sites of granulopoiesis in the splenic red pulp (Wolf and Trentin, 1968). They have a short life span (6—11 days), are found in the blood in large numbers compared with lymphocytes ($3-6 \times 10^3$ ml^{-1} : 1.5×10^3 ml^{-1}), and have the property of leaving the blood vessels and invading areas of inflammation (see Florey, 1970; Grant, 1973). During an inflammatory process granulocytes penetrate the regional lymphatic circulation that leads them to the regional lymph node (de Sousa and Parrott, 1969; Herman *et al.*, 1972) but in the mammal they are not found in central lymph. Interest in the process of invasion of sites of inflammation developed from the observations, descriptions, and drawings of that most thorough and competent late nineteenth century generation of pathologists that included such remarkable men as Addison, Virchow, Conheim, and others (see Florey, 1958).

From those early careful drawings of inflamed frog tongues and frog foot webs examined under unsophisticated microscopes, we learnt that which 100 years later we still go on teaching and reviewing to our shame and their credit: the rate of blood flow in an inflamed or injured area increases during the early period of inflammation (30 min) and a dilatation of the smaller vessels is observed. With time, the flow rate decreases in spite of the persisting vasodilatation, and leucocytes start to

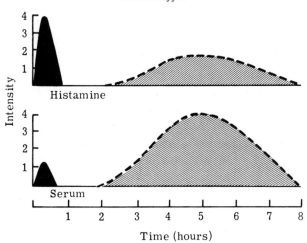

Fig. 3.7 Diagram illustrating the lack of correlation between intensity of vascular permeability (■) and leucocyte emigration (▨) to skin sites where histamine or serum was injected. Original data from Hurley (1963), diagram modified from Grant (1973).

adhere to the wall of small vessels which eventually they cross into the surrounding tissues. With the advent of microcinematography and electron microscopy, knowledge of this process has gained considerable numerical and ultrastructural detail, but little gain in understanding of mechanism (see Grant, 1973).

Granulocyte rolling along a vessel wall is proportional to the velocity of the blood flow in venules where the velocity is in the range 300–1000 μm s^{-1} (Atherton and Born, 1973). Sticking to the endothelial wall is clearly influenced by a lesion of the endothelial wall (Baez and Kochen, 1965; Grant, 1973). Although the exact nature of this lesion remains unknown, migration of granulocytes from an injured vessel can be dissociated from altered permeability of the vessel, i.e the presence of polymorphs does not influence permeability (Humphrey, 1955) and increased vascular permeability does not necessarily result in a proportional increase in leucocyte migration to the area (Hurley, 1963; Logan and Wilhelm, 1963). This latter point merits analysis in detail.

Humphrey (1955), in a study of vascular permeability in animals depleted of polymorphs, demonstrated that the early increase in permeability observed during an Arthus reaction was unaffected by the absence of polymorphs. Hurley (1963), in a study of the effect of the injection of a number of substances into the rat skin, comparing vascular permeability and granulocyte migration, was able to dissect patterns of polymorph emigration unrelated to degree of vascular permeability (Fig. 3.7). For example, after the administration of histamine, saline, and homologous serum, leucocyte migration occurred only several hours (2–3) later, being maximal at 5–6 h after injection. In spite of a considerable increase of vascular permeability observed with histamine, only a moderate degree of extravasation was seen compared with the serum injection. Injection extracts of burned skin, extracts of polymorphonuclear cells, and of serum after incubation *in vitro* with other tissues, resulted in a massive migration of cells into the injected area within half an hour.

The results of this type of experiment underline once more the wide gap existing between traffic of inert particles and traffic of cells. The materials that induced the massive migration of cells from the blood vessel are all known to be chemotactic for polymorphs *in vitro* (see Wilkinson, 1976). Thus, cell traffic cannot be visualized exclusively as a biophysical problem; the fact that cells are not deflected by increased vascular permeability from their well-defined route, but can leave it rapidly for 'alarming' molecular reasons, makes them look more like people rushing into the scene of a car crash than balls rolling along a bowling alley.

3.4 CELL TRAFFIC WITHIN THE LYMPHOID SYSTEM

'It must now be clear that not only are the fate and total life span of the blood lymphocytes unknown, but that there is, as yet, no clue to their possible function in the body. The classical method of revealing such a function would be to remove all the lymphocyte-producing tissue from an animal and observe the effect. It is, of course, impossible to remove surgically all the tissue owing to its scattered distribution'. (Florey and Gowans, 1958).

Unlike the unchanged descriptions of the cellular events of inflammation, our basic views of function and arrangement of the lymphoid system are radically different from those held not 100 but 20 years ago. In the nineteen-fifties, the dominating view of the lymphoid system was still that of Virchow (1858) who postulated that the lymph nodes, the spleen, and the Peyer's patches acted as centres of production of lymphocytes which entered the lymph and from the lymph entered the blood. The lymphocyte journey then was thought to be a short one, of a short-lived cell with no obvious function to justify its intriguing existence. The classical experiment by Gowans (1957) showing that reintroduction of drained thoracic duct lymphocytes into the blood circulation of the rat was the only procedure that successfully kept thoracic duct output of lymphocytes constant over a long period of drainage (more than 2 days) opened a new window into knowledge of cell traffic. After this series of elegant experiments (Gowans, 1957, 1959, 1966; Gowans and Knight, 1964), lymphocytes, believed in the Virchowian tradition to be made in the lymph nodes, drained into the lymph, entered the blood, and *recirculated* back into the lymph (Gowans, 1959, 1966).

It was from the almost simultaneous opening of so many other windows and doors in the 1956–1966 decade that the strongly built Virchowian view of the lymphoid system finally fell apart. In all probability that decade will be described by historians of science as an 'epidemic' of understanding of lymphocyte function, rather like a creative and intellectually procreative form of encephalites. An infectious interest in immunological function swept across the world from Australia, to England, to Connecticut and Minnesota in the United States, across species from the chicken (Glick *et al.*, 1956; Simonsen, 1962; Warner and Szenberg, 1962; Warner *et al.*, 1962; Cooper *et al.*, 1966), to the mouse (Medawar, 1958; Miller, 1961, 1962; Parrott, 1962; Humphrey *et al.*, 1964; Parrott *et al.*, 1966; Dalmasso *et al.*, 1962, 1963; Ford, 1966; Davies *et al.*, 1966), to the rat (Gowans, 1962, 1965; Waksman *et al.*, 1962), across the board of scientific activity from experimental work (all the above) to theory (Jerne, 1955; Burnet, 1957).

The views of the lymphoid system that presently we, and our textbooks (Humphrey and White, 1970; Roitt, 1974; Good and Fischer, 1971; Greaves *et al.*, 1974) hold, have their foundations in that very remarkable decade. It is outside the scope of a review to refer in detail to widely published textbook material. Briefly, multipotential stem cells in the haemopoietic pool of the foetal liver (Owen *et al.*, 1974) or the adult bone marrow (Ford, 1966; Davies *et al.*, 1967) give rise to cells that can differentiate into two functionally different classes of lympho-cytes, depending on their journey through the thymus. Those that do travel through the thymus are influenced within it by thymus specific or ubiquitous products (Goldstein *et al.*, 1975; van Bekkum, 1975) entering the blood circulation with a set of surface antigens specific for thymus-derived cells (Williams *et al.*, 1971; Schlesinger, 1970; Komuro and Boyse, 1973; Cantor and Boyse, 1975a, 1975b) virtually undetectable amounts of surface immunoglobulin, capable of responding by division to antigen but unable to produce antibody. Those that do not circulate through the thymus differentiate directly into a population characterized by the presence of immunoglobulin on their surface and ultimately capable of producing antibody. Production of normal amounts of antibody to most antigens however, requires the presence and cooperation of both B (non-thymus-derived) and T (thymus-derived) cells.

Circulating lymphocytes, therefore, are not only the product of cell division in lymph nodes as described by the early pathologists, but the long lived progeny of cells that travelled a long distance before getting there, a journey that started in the blood circuit, passes through the peripheral lymphoid organs and leads into the lymph (Pearson *et al.*, 1976). From the lymph, lymphocytes *recirculate* into the blood, not the other way around.

The questions of life span, route of circulation, significance of lymphocyte traffic for the immune response, distribution within the lymphoid organs while in transit, have been reviewed in detail over the last ten years (Gowans, 1966; Ford and Gowans, 1969; Parrott and de Sousa, 1971; de Sousa, 1973; Ford, 1975; Sprent, 1976b). In this review, I shall ask myself slightly different questions:
(1) How much do we know of bone marrow cell traffic?
(2) Why do thymus, spleen, lymph node and thoracic duct lymphocytes have different blood to lymph velocities?
(3) What factors influence lymphocyte velocity?
(4) Do T and B cells meet?

3.4.1 How much do we know about bone marrow cell traffic?

Insight into the traffic of bone marrow cells derived originally from work on the efficacy of different cell inocula in reconstituting lethally irradiated animals (Micklem and Loutit, 1966). Administration of small cell doses in the mouse (Till and McCulloch, 1961) though inadequate as a replacement measure, proved to be one of the most useful tools in the understanding of bone marrow cell differentiation (Trentin *et al.*, 1971). Nine days after the intravenous injection of $4-5 \times 10^4$ bone marrow cells, well defined colonies develop in the spleen which can be seen and counted macroscopically. Each spleen colony is known to derive from one single uncommitted multipotential cell (CFU, Colony Forming Unit) which will give rise to one of four haemopoietic lines: erythroid, neutrophilic granuloid, eosinophilic granuloid or megakariocytic,

ECOTAXIS

(spleen)

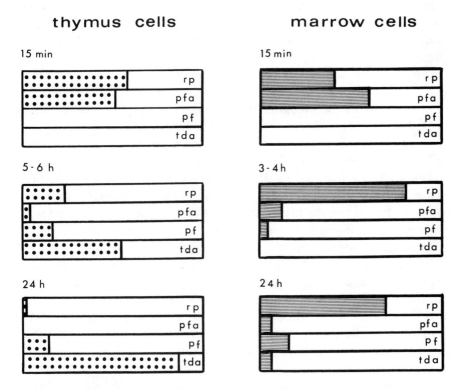

Fig. 3.8 Differential migration of ^{3}H-adenosine labelled thymus and bone
marrow cells in the mouse spleen (ecotaxis). At early times after injection
both cell types are present in the red pulp (rp) and perifollicular area (pfa).
At 24 h the majority of thymus cells are found in the thymus-dependent area
(tda) whereas the bone marrow cells stay in the red pulp. Data from
de Sousa, (1971).

depending on the nature of the stroma with which it interacts (Wolf and
Trentin, 1968).

Wolf and Trentin's experiments (demonstrating that haemopoietic
stem cells, placed in bone marrow stroma implanted in the spleen, differentiate along the line (granuloid) which is the predominant one in
bone marrow and that the same multipotential cell growing in the spleen
stroma gives rise mostly to erythroid cells) are, in my opinion, the most
elegant and convincing experiments of how the site where an uncommitted

cell finds itself influences its fate. We have seen that after an intravenous injection, cells must pass the pulmonary circuit. Red cells do this and then continue through spleen, etc. to be found always in greatest quantity in the blood itself. Bone marrow cells obviously do not do this. They pass the pulmonary and hepatic circuits shortly after injection, but their alternative destination, in addition to the recipient's marrow itself, is the splenic red pulp (Fig. 3.8). This is most striking in the irradiated recipients because the cells settle and proliferate, but the distribution pattern in intact syngeneic recipients is identical. Only a small proportion of cells in a bone marrow cell inoculum ever reaches the lymph nodes (de Sousa, 1971; Isles, 1974; Schlesinger and Israel, 1974).

There is one other point derived from the aforementioned experiments. Differentiation is influenced by the nature of the stroma, but the rate of proliferation is influenced by quantitative as well as qualitative factors, by the numbers as well as the type of neighbours a travelling cell encounters. In spite of the identical distribution of the inoculated bone marrow in intact syngeneic recipients and of small donor-derived colonies no macroscopic or microscopic colonies of the type observed in irradiated recipients are ever found in a recipient with its full normal cell quota.

Finally, a donor bone marrow cell sub-population enters the thymus and therein differentiates along a lymphoid cell line. The exact nature of this sub-population, whether it is a multipotential cell like the ones found in the spleen, or already a specific precursor committed to differentiate in a thymus, remains elusive (Roelants *et al.*, 1975). Judging from the radioactivity recovered in the various organs after an intravenous injection of ^{51}Cr-labelled bone marrow cells, only a negligible proportion of the cells (0.3%) enters the thymus (Isles, 1974). Moreover, attempts to recover CFU from the thymus at intervals ranging from 15 min to 15 days (Table 3.4) failed to reveal the presence of multipotential cells in any number (Trentin *et al.*, 1971). It is likely that the nature of the interaction between the bone marrow cell and the thymus stroma is very different from the one with the splenic or the bone marrow stroma, involving adenyl cyclase activation (Singh, 1975; Scheid *et al.*, 1973).

3.4.2 Why do thymus, spleen, lymph node, and thoracic duct lymphocytes have different blood to lymph velocities?

A multipotential or committed precursor haemopoietic cell, having made what still remains a mysterious journey within the thymus, gives

Table 3.4 Recoverability of CFU from the thymus of irradiated mice*

Time after injection	No. of CFU recovered/one thymus
15 min	2.8
30 min	1.7
1 h	0.4
2 h	0.8
3 h	1.2
4 h	0.7
5 h	0.3
1 day	0
4 days	0
6 days	0
9 days	0
11 days	0
15 days	0

* Data from Trentin *et al.*, 1971.
  Whole body irradiated (1, 100 rad) (C57 x A) mice received an injection of
  5×10^6 syngeneic bone marrow cells.

rise to a progeny of cells destined to leave it, entering the blood circulation (de Sousa, 1973; Levey and Burleson, 1975). In the chicken,
B lymphocytes leave their well-defined site of origin, the bursa of
Fabricius, enter the circulation and distribute themselves in clearly
defined splenic areas (Fig. 3.9) (Durkin and Thorbecke, 1973; de Sousa,
1973). In the mammal, traffic of B cells from the foetal liver, where
their development seems to take place first in ontogeny (Owen *et al.*,
1974) to the sites where they are found in the peripheral lymphoid
organs (de Sousa *et al.*, 1973) has not been defined.

If one makes a thymus cell suspension, labels it *in vitro* and introduces it in a syngeneic blood circulation, 1 h later, very few cells will
have travelled all the way to the lymph nodes, even at 24 h, only a small
proportion will be found there, compared with the fate of a lymph node
or a thoracic duct lymphocyte suspension (Table 3.5). If instead of
taking the lymph node or the thoracic duct lymph as our point of
reference, we take the spleen, (as we saw already a considerable proportion of bone marrow cells mitigate to the spleen,) a high proportion
(40%) of a spleen cell inoculum is also found in the spleen, and a remarkably constant proportion (around 15–20%) of thymus, lymph node
and thoracic duct lymphocytes is found in the spleen at 24 h after

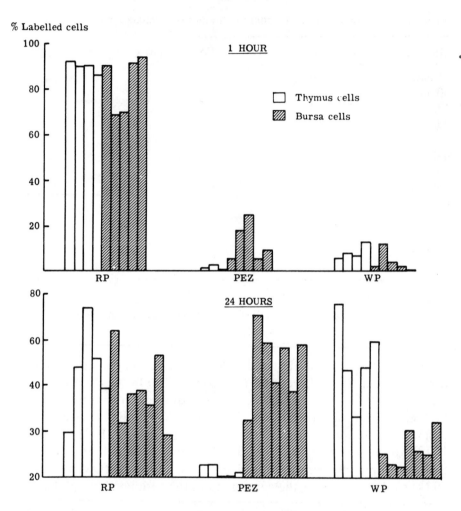

Fig. 3.9 Differential migration of [3] H-adenosine labelled autologous thymus and bursa cells in the chicken spleen (ecotaxis). At 1 h after injection both cell types are found in the red pulp (rp). At 24 h, however, most thymus cells are present in the white pulp (wp) round the central arteriole, whereas the bursa cells remain in the red pulp and round the ellipsoids, in the periellipsoidal zone (pez). Bursa cells are also found in germinal centres (not represented). Data from de Sousa, Eslami and White, (unpublished experiments). Each bar represents one animal.

intravenous injection (i.v.) The result of various groups working on lymphocyte traffic are so remarkably similar (Lance and Taub, 1969; Zatz *et al.,* 1972; Sprent, 1974; Isles, 1974; Freitas and de Sousa, 1975) that they can be expressed as pooled spleen: lymph node ratios (Table 3.6).

Expressing cell velocity in this way, it becomes apparent that the population that travels from the blood to the lymph node fastest is the one from thoracic duct lymph (ratio = 1) followed by the lymph node cell population (1.5), then the spleen (4), the bone marrow (6), and the thymus (15), assuming that the percentage recovered in the spleen is fairly constant. This differential quantitative lodging of lymphoid cell populations to the spleen and lymph nodes is now known in the literature as the 'spleen-seeking' and 'lymph node seeking' phenomenon, reflecting the existence of 'spleen-seeking' and 'lymph node seeking' populations. The use and adoption of these expressions is most revealing in demonstrating how much the interpretation of one's own results is influenced by the historical development of the subject. When, in the field of red cell traffic it was discovered that antibody treated red cells were sequestered in the spleen, nobody thought of calling these modified cells 'spleen-seeking', they were envisaged more like 'spleen-caught' (or sought by the spleen). Yet, the basic mechanism of distribution of lymphocytes and red cells in the spleen must be the same, i.e. traffic of a cell through an organ is influenced by interaction of that cell with the cells it meets on its way, between arterial and venous blood.

The expression 'spleen-seeking' taken to its logical conclusion means that if the spleen was situated in an arterial-venous circuit in a toe, lymphocytes would find their way to the toe. In all probability, the only reason why some lymphocytes are found in greater numbers in the spleen at a given time after injection is simply because the spleen is 'in their way' and they have a tempo of circulation which is slower than the other population.

The fact that lymphocyte populations differ in their circulation tempo is now well established from experiments using pure populations of labelled T or B lymphocytes (Howard, 1972; Sprent, 1974; Freitas and de Sousa, 1975). T lymphocyte traffic is faster than B lymphocyte traffic. If one follows the fate of T and B cells at different times after intra-

Table 3.5 Differential circulation velocities† of cell populations within the lymphoid system expressed as spleen/lymph node ratios‡

Cell population	Spleen/lymph node ratio
Thymus	15
Bone marrow	6
Spleen	4
Lymph node	1.5
Thoracic duct	1

† Pooled data from: Lance and Taub, 1969; Zatz *et al.*, 1972; Sprent, 1974; Isles, 1974; Freitas and de Sousa, 1975.

‡ % injected radioactivity measured at 24 h.

Table 3.6 Summary of published work on the effect of different treatments of labelled lymphocytes *in vitro* on their *in v* fate following intravenous injection

Reference	Inoculum	*In vitro* treatment	Distribution change				
			Blood	Lungs	Liver	Spleen	Lymph node
Gesner and Ginsburg 1964	Rat TDL	Glycosidases	ND	↔	↑	↓	↓
Berney and Gesner 1970	Thymus	Trypsin	ND		↑	↑	↓
Woodruff and Gesner 1968	Rat TDL	Trypsin	ND	↔	↑	↑	↓
	Rat TDL	Neuraminidase	ND	↔	↑	↓	↓.
		Heating	ND	ND		ND	↓
Martin, 1969	Mouse thymus		↔	↔	↑	↓	↓
	LNL:TDL	ALS	↔	↔	↑	↓	↓
Taub et al., 1972	Mouse LNL	B. pertussis Supernatants	↔	ND	↗	↗	↓
Zatz et al., 1972	Mouse LNL	NaIO₄ oxidation	ND	ND	↑	↓	↓
Gillette et al., 1973	Mouse LNL	Con-A	ND	ND	↔	↑	↓
Schlesinger and Israel, 1974	Mouse LNL	Con-A	ND	ND	↗	↙	↓
	Mouse LNL	PHA	ND	ND	↔	↓	↓
	Mouse	Con-A	ND	ND	↔	↓	↓
	Thymus	PHA	ND	ND	↔	↓	↓
Woodruff, 1974	Rat TDL	Trypsin	ND	ND	↔	↔	↓
		Puromycin	ND	ND	↑	↓	↓
		Trypsin + Puromycin	ND	ND	↑	↓	↓
Freitas and de Sousa, 1975	Mouse LNL	Con-A	ND	↔	↔	↑	↓
	Mouse LNL	PHA	ND	↔	↑	↑	↓
	Mouse LNL	LPS	↔	↔	↔	↑	↓
	Mouse LNL	PL-A	↔	↑	↑	↔	↓
		Tryspin	↔	↔	↑	↑	↓
		Neuraminidase	↔	↓	↑	↓	↓

↔ No change; ↑ : increase; ↓ : decrease; ↗ : slight increase; ↙ : slight decrease; LNL : Lymph node lymphocytes; TDL: thoracic duct lymphocytes; ND: not done.

venous injection, this differential velocity is first apparent in the lungs, where at 1 h B cells are present in significantly higher numbers (27% of injected radioactivity) than T cells (12% of injected radioactivity). Another important finding regarding tempo of circulation of lymphocytes is the differential velocity of migration to the thoracic duct lymph of 'virgin' (unprimed) and primed B cells (Strober, 1972; Strober and Dilley, 1973). After priming with antigen, B cells acquire the capacity to leave the blood circuit and circulate in the lymph circuit more readily.

Clearly the differential blood to lymph velocity of lymphocyte sub-populations depends on their sojourn in the organs intercalated in the pulmonary and systemic blood circulations, of those in the systemic circulation the most important are the liver and the peripheral lymphoid organs namely the spleen, lymph nodes and Peyer's patches in the mammal, the spleen in the chicken (Durkin and Thorbecke, 1973) and the spleen and the kidney in fish (Ellis and de Sousa, 1974). Whereabouts in these organs are the cells found?

As one might expect from the earlier descriptions of the behaviour of cells in flows (Section 1) and of the geometry of the circuits (Section 2), in the liver and lungs, lymphocytes are found in the sinusoids and alveolar capillaries respectively. In the spleen, however, lymphocytes are only found in the rich capillary network of the red pulp, in any of the species studied, at the early times after intravenous injection (up to 3 h). After this lymphocytes migrate into the white pulp, bone marrow cells within the red pulp cords, and within the white pulp, lymphocytes do the equivalent to *in vitro* cell 'sorting out' (Curtis, 1973) and thus 'sort themselves out' into two major areas, one of which is occupied by the B cells in the periphery of the Malpighian body in the mammal, or within the germinal centre and round the ellipsoid in the chicken, and another at the centre, immediately round the central arteriole, occupied by the T cells, both in the mammal and in the chicken (de Sousa, 1973).

The major route of lymphocyte entry into the lymph node from the blood is not the capillary network but the section of the post-capillary venule trunk situated in the mid cortex. In lymph nodes from intact animals this mid-cortical section of the post-capillary venule is character-ized by a singular high endothelium of cuboidal cells. In animals depleted of T cells, however, the post-capillary venule endothelium in this region is as flat as elsewhere in the node. After utilizing this apparent common pathway of entry into the lymph node, lymphocytes within the node also 'sort themselves out'. B lymphocytes are found in the peripheral cortical nodules, whereas T lymphocytes remain in the mid cortex.

Specific lymphocyte migration and lodging in separate areas, is also seen in the Peyer's patches (de Sousa, 1973; Parrott and Ferguson; 1974; Parrott, 1976).

The ability of mature lymphocytes to migrate and arrange themselves in specific sites of the peripheral lymphoid organs was called ecotaxis by de Sousa (1971) a term which, for its lack of obvious operational meaning, has not been adopted by others.

Ecotaxis was thought at one time to define the capacity of a cell 'to make the decision' to go to a site, rather as 'spleen-seeking' and 'lymph node seeking' imply. However, all the evidence from *in vivo* and *in vitro* work on modulation of lymphocyte traffic (see Section 3.4.3.) and on adhesive interactions between T and B lymphocytes, point to the fact that the ultimate specific position occupied by a T or a B cell in a lymphoid organ must represent the nature of the interaction of that cell with its neighbours (de Sousa, 1976a; Freitas and de Sousa, 1975, 1976a, 1976b, 1976c). Thus, this interaction must control not only positioning but also velocity of traffic , whereas in the lungs, liver and spleen red pulp, the major control must reside in the interaction between circulating cells and capillary walls. Within the spleen white pulp, the lymph node and the Peyer's patches, control of traffic must result mostly from interaction between lymphocytes themselves.

3.4.3 What factors influence lymphocyte traffic velocity?

I shall divide the answer to this question into two further subsections: (a) factors that modify the lymphocyte itself (b) factors that modify the host in which lymphocytes circulate.

3.4.3.1 Factors modifying the circulating lymphocyte in vitro
We saw that the aim of most experiments on red cell traffic was determined by preoccupation with the mechanism of anaemia. The aims of most experiments on modulation of lymphocyte traffic altering the lymphocyte itself, were predetermined by the original emphasis put on the interaction between post-capillary venule and lymphocytes in lymph nodes (Gowans and Knight, 1964; Marchesi and Gowans, 1964). Interest in the molecular nature of this interaction led Gesner (1966) to postulate that sugars on the surface of the lymphocyte might play a role in recognition of the endothelium. Had someone postulated a role for anti-lymphocyte antibodies (as they did for red cells) the kind of experiments that could have followed, would have been very different indeed.

As it happened, the majority of experiments modifying lymphocytes *in vitro*, consist of treatments designed to modify surface carbohydrates (Gesner and Ginsburg, 1964; Woodruff and Gesner, 1969; Woodruff, 1974; Zatz *et al.*, 1972; Gillette *et al.*, 1973; Taub, 1974; Schlesinger and Israel; Freitas and de Sousa, 1975) with a few notable exceptions: the experiments of Morse's group (Morse, 1964; Morse and Riester, 1967a, b; Morse and Bray, 1969; Morse and Barron, 1970; Taub *et al.*, 1972) on the mechanism of the lymphocytosis induced by *B. pertussis*: Durkin, Caporale and Thorbecke's experiments on migratory patterns of B lymphocytes modified by anti-immunoglobulin (Durkin *et al.*, 1975) and Martin's work (1969) on the effect of anti-lymphocyte serum.

A summary of the work of various groups on the effect of treatment *in vitro* with various enzymes, lectins, *B. pertussis* cultures, supernatants etc. is included in Table 3.6. On the whole, *any* pretreatment of labelled lymphocytes *in vitro*, alter their overall pattern of distribution on transfer, after intravenous injection. The constant feature of this alteration is decreased lymph node entry; variable changes in distribution in other organs occur concurrently and deserve to be looked at in some detail.

Neuraminidase treatment results in a transient increase in numbers of cells in the liver (Gesner and Ginsburg, 1964; Woodruff and Gesner; 1969, Freitas and de Sousa, 1976a) which is no longer apparent by 24 h; treatment with trypsin causes a transient increase in liver and spleen which also is not apparent at 24 h, concanavalin A has a similar effect i.e. increased initial localization in the spleen (Gillette *et al.*, 1973; Freitas and de Sousa, 1975), although this effect does not always seem to be reproducible (Schlesinger and Israel, 1974). One intriguing effect is that of phospholipase A for causing an early (1 h) accumulation of lymphocytes in the lungs (Freitas and de Sousa, 1976a, 1976c).

Therefore the decrease in lymphocyte traffic into the node can be visualized as the result of modification of the interaction of the circulating cells with endothelial and other cells elsewhere in the blood circuit. Why cleaving the terminal sialic acid in a glycoprotein modifies its interaction with the liver (Morell *et al.*, 1971) and cleaving it on a lymphocyte surface alters the cell's interaction with the endothelium of the liver sinusoid, binding concanavalin A alters the interaction with cells in the liver sinusoids and in spleen, and phospholipase A affects the alveolar cell-lymphocyte interaction, is the pertinent question. Concanavalin A also modifies the interaction of the lymphocyte with the lymph node post-capillary venule, unusually high numbers of labelled cells are seen 'stuck' to the endothelial cells without entering the node,

Table 3.7 Effect of splenectomy on the distribution of radioactivity from
^{51}Cr-labelled Con-A, LPS or trypsin treated mouse lymph node cells after
intravenous injection

Compartment	Treatment		
	Con-A	LPS	Trypsin
Blood	ND	ND	↑
Lymph node	↔	↔	↓
Liver	↔	↔	↔
Lungs	↔	↔	↔

ND: not done; ↔ : not different from control cells; ↑ : higher than control cells;
↓ : lower than control cells.

but that must be seen just as part of the general modification of the
lymphocyte-endothelium interaction which results in its temporary
sequestration in the spleen.

In order to demonstrate that decreased lymph node entry is nothing
uniquely important but the reflection of other equally important events
occurring elsewhere (de Sousa, 1976a), we have done a series of tracing
experiments in mice that had their spleens removed before cell transfer
(Freitas and de Sousa, 1975, 1976a, b, c). If our view was correct, in
those cases with temporary increased sequestration in the spleen,
splenectomy would enable the transfused labelled cells to reach the
lymph nodes at the same speed as that of control untreated cells
(Table 3.7). This prediction proved correct for Concanavalin A and
LPS treatment, after treatment with trypsin, however, the labelled
lymph node cells in the splenectomised recipients stayed in the blood
in much higher quantities than the control, untreated cells. This effect
of trypsin could be the result of removal of a specific receptor for the
lymph node post-capillary venule endothelium (Woodruff, 1974) but a
more simple explanation is the more generalized action of trypsin on
the cell surface, i.e. reduction in adhesiveness (Curtis, 1973). It is of
interest to note that the effect of supernatants of cultures of *B. pertussis*
is similar to that of trypsin in splenectomised mice, i.e. decreased lymph
node entry occurs with a concurrent increase of radioactivity in blood,
although the nature of the active component in this case has not been
defined*.

* Note added in proof: Leukocytosis promoting factor (LPF) has now been isolated
  by detailed physical, chemical, and electron microscopical analysis (Morse and

3.4.3.2 Influence of factors acting in vivo

Most of the work in this field is closely linked to interest in the development of the immune response; thus, of the factors modifying the circulation of lymphocytes *in vivo*, antigen has been the one most thoroughly investigated. Antigens can be presented as separate entities from circulating labelled lymphocytes or as integral components of the lymphocyte itself in the form of histocompatibility antigenic determinants (for review, see Ford, 1975; and more recently Frost *et al.,* 1975; Cahill *et al.,* 1976; Sprent, 1976).

It was first observed by Hall and Morris (1965), who called it the 'shut-down' phenomenon that after the local administration of antigen there is a transient reduction in the numbers of lymphocytes leaving the draining lymph node via the efferent lymphatic. Later it was demonstrated that both the administration of adjuvant materials (Dresser *et al.,* 1970) and of antigen (Zatz and Lance, 1971) result in what was thought to be an obligatory sequestration of circulating cells and accordingly called lymphocyte trapping (Zatz and Lance, 1970). The recent work of Cahill *et al.,* (1976) on the transit time of labelled lymphocytes through lymph nodes undergoing immune responses to different antigens, however, has shown that the migration of the bulk of ^{51}Cr-labelled lymphocytes through a lymph node draining antigen does not differ greatly from the migration time through non-draining nodes. The constant difference resides in the actual absolute numbers of lymphocytes entering an antigen draining node, a result to be expected from the microcirculation studies of Herman and his colleagues (Herman *et al.,* 1972) described earlier (Section 3.2). After the administration of influenza virus, Cahill *et al.,* (1976) found that in addition to the increased input into the antigen stimulated node, circulating lymphocytes were delayed in their passage through the node. The exact cellular basis of this delay is not fully understood but present evidence indicates that modified interaction between macrophages and lymphocytes may be the cause of the lymphocyte delay (Frost, 1974; Mongini and Rosenberg, 1976).

One other series of experiments on modified lymphocyte traffic

(continued)

Morse, (1976), *J. Exp. Med.* **143** 1483. It contains 14.5% nitrogen, it is lipid- and carbohydrate-free, and it seems to be composed of four polypeptide subunits with a probable minimum mol. wt. of 86 000. The possible mechanism of action of the purified component raises for the first time the possibility that cyclic nucleotides may play a role in the control of lymphocyte traffic.

in vivo that has escaped attention recently but in which the importance of short-range cell interaction in the control of lymphocyte traffic is also apparent, is that of Morse's group in the effect of *B. pertussis* (Morse, 1964; Morse and Riester, 1967a, 1967b; Morse and Bray, 1969; Morse and Barron, 1970; Taub *et al.*, 1972).

B. pertussis causes a striking blood lymphocytosis in mice. Morse demonstrated that the increase in numbers of lymphocytes was not due to newly formed cells but to the retention of long lived cells in the blood (Morse and Riester, 1967a). Secondly, it was shown that the lymphocytosis was confined to the blood and that lymphocytes failed to circulate to lymph (Morse and Riester, 1967b). Finally, in a most thorough comparative study (Taub *et al.*, 1972) of the effect of the whole organism versus a 'lymphocytosis promotor factor' obtained from supernatants of *B. pertussis* cultures on traffic of injected ^{51}Cr-labelled lymph node or thoracic duct cells *in vivo*, the following results were obtained:

(1) Administration of the supernatant before or after the labelled cells resulted in their redistribution in the host: more radioactivity was recovered from the blood, less from the lymph nodes and less from the spleen.

(2) Administration of the whole organism provoked a similar decrease in lymph node entry and blood increase, but, in addition, increased numbers of cells were retained in the spleen (Table 3.8).

(3) *In vitro* treatment of labelled cells with plasma, from *B. pertussis* supernatant injected mice, failed to alter their migration.

(4) Unlabelled erythrocytes or lymphocytes which had bound *B. pertussis* supernatant *in vitro*, when mixed with untreated labelled cells, altered their migration.

The implications of these observations are manifold, two aspects in particular must be stressed. Firstly, the fact that cells, not plasma, carried on their surface a factor that after elution could actively influence the migration of other cells, indicating that direct or short range cell interactions are at play in this system. Secondly, the fact that after an intravenous injection, the whole organism, but not the supernatant, caused some degree of cell retention in the spleen, an additional illustration of lymphocyte trapping. Why the organ-seeking terminology was not applied in this case and the term 'trapping' started to be used is unexplained (Zatz and Lance, 1971). In essence 'lymphocyte trapping' is as much 'antigen-draining-organ-seeking' as anything else and 'antigen-draining-organ-seeking' is as much the result of an interaction between a macrophage and a lymphocyte (Frost, 1974; Mongini and Rosenberg,

Table 3.8 Distribution of ^{51}Cr-labelled normal, syngeneic/lymph node cells in mice treated with *B. pertussis* organism or culture supernatant 3 days before cell transfer*

Pretreatment	WBC	% injected radiactivity				
		Blood	PLN	MLN	Spleen	Liver
Intact organism	45 400	2.4	3.5	6.0	21.4	16.8
Supernatant	25 300	3.4	3.9	5.8	6.6	18.6
Untreated	4 100	0.7	6.0	11.9	12.7	15.4

* Data from Taub *et al.*, 1972; PLN: peripheral lymph nodes; MLN: mesenteric lymph node.

1976), or two lymphocytes, or a red cell and a lymphocyte (as suggested by Taub *et al's* results) as the original 'spleen-seeking' term failed to imply.

Do lymphocytes specifically sensitized to an antigen 'seek' that antigen outside the main route of their everyday traffic? There is no clear cut demonstration that they do.

Most of the early experiments were done with labels that did not select for dividing cell populations. When this is done, however, some degree of site-specific migration is observed (Griscelli *et al.*, 1969; Parrott and Ferguson, 1974; Parrott *et al.*, 1976; Hall, 1976). One intriguing aspect of this specificity is that it seems to be related to the site where the antigen is applied and not to the antigen itself. For instance, [125]IUdR labelled blast cells, obtained from lymph nodes draining the mouse skin painted with a sensitizing agent, on transfer to challenged syngeneic recipients, migrate within a short time to skin sites challenged either with that or any other agent. [125]IUdR labelled blast cells transferred into animals undergoing a gut infection and not skin challenged, however, only penetrate the gut in negligible amounts. [125]IUdR mesenteric labelled lymph node cells, on the other hand, migrate readily to the gut and not to the skin (Parrott *et al.*, 1975; Rose *et al.*, 1976).

These observations, in essence similar to those first described by Griscelli *et al.*, (1969) on the fate of mesenteric and peripheral lymph node cells migrating to the gut, now confirmed by the recent work of Hall (Hall, 1976) leave us with the question as to whether in each group of lymph nodes there is a small population of lymphocytes committed to tissue antigen specificities of the area it drains.

One other group of recent experiments of paramount relevance to

understanding host control of lymphocyte traffic, are the experiments
of Bell and Shand (1975) based on the Dresser (1961) and Celada (1966)
models of adoptive transfer of immunity in irradiated recipients.
Thoracic duct lymphocytes from animals immunized to HSA, transferred
immunity successfully in irradiated but not intact recipients. Both the
magnitude and the affinity of antibody were greater in the irradiated
recipients and this was related to an impaired blood-lymph circulation
of the labelled immune cells. Reduction in recovered radioactivity from
the recipients' thoracic duct lymph concurred with increased recoveries
in the lymphoid organs. Moreover, administration of unlabelled non-
immune thoracic duct lymphocytes, before the injection of the labelled
cells corrected the impairment in lymphocyte circulation, and functionally, had the effect of suppressing the adoptively transferred response.
These results stress once more the fact that lymphocyte interactions
are at play in controlling rate of cell traffic, but, in addition, they have
the merit of directly relating control of lymphocyte traffic to control of
antibody production.

Antigen, by increasing the input of recirculating lymphocytes into
the draining lymphoid organ and in some instances slowing down their
output, at least in theory, must create ideal physical conditions for the
occurrence of direct cell-cell interactions known to be essential for the
development of an immune response.

This leads us to the next question.

3.4.4 Do T and B cells meet *in vivo* ?

How, where and when can two cells travelling at different speeds, only
transiently utilizing common routes of traffic in the red pulp and the
perifollocular area or marginal zone of the spleen, occupying different
areas within the white pulp and the lymph nodes, ever find the opportunity to interact and do all the things that our neat, static diagrams predict?

Throughout this review it has become apparent that cells seem to
behave more as simple minded people would expect of people, than
great minds expect of spheres (Einstein, 1906). If a man running alone in
a single track at high speed had to pass on a message to another man
running at a lower speed in the adjacent track, he would probably, do
one of two things: slow down, or, leave the message with a third person
in a fixed position, who could then pass it on to the slow runner when
he came. But cells do not run singly in lonely tracks. Cell traffic is a
crowded affair more like rush hour traffic in and out of a railway station.

If a man and a woman in a railway station suddenly separated by the avalanche of people pouring out of an arriving train, decide to say how much they mean to each other, they too will have to do one of two things: to shout (a pretty useless and unbecoming way of passing on an essentially delicate message) or wait till the moving crowd finds its way out of the station. Having passed their message, it is unlikely that the avalanches of people from later trains will ever separate them again.

How far from these analogies is that which we already know of lymphoid cell traffic and lymphoid cell interactions during an immune response?

(1) *Slowing down* of traffic occurs as described in the previous section (see Ford, 1975).

(2) *Passing of a message to a third fixed cell* has been demonstrated in a series of elegant studies by Feldman and co-workers, (Feldman, 1972; Feldman and Basten, 1972a, 1972b; Feldman *et al.,* 1973) indicating that activated T cells release factors which bind to the surface of macrophages whereupon B cells can find it and be triggered to the production of antibody. The efficiency of other T factors that do not require the macrophage step, and operate at short range from T cells directly to B cells (see Munro and Taussig, 1975) is likely to increase in the situation of slowed down traffic (see also below).

(3) *Being separated by a crowd* of unlike cells. From experiments on the nature of lymphocyte adhesive interactions (Curtis and de Sousa, 1973, 1975) it has been shown that *in vitro*, T and B cells, or factors of low molecular weight released in their respective supernatants, diminish the adhesion of the unlike cell type (Fig. 3.10). T cell factors diminish the adhesion of B cells and vice versa. Extrapolating from these observations to the *in vivo* situation, T and B cells would be expected to occupy different positions in the lymphoid organs, as indeed they do, and the event of two unlike cells being found physically together, is an unlikely one. I attempted to visualize this by developing a double layer autoradiography technique (Shand and de Sousa, 1974) which enabled me to trace *in vivo* the fate of ^{3}H-adenosine labelled T cells and ^{14}C-labelled bone marrow cells using a classical thymus—marrow co-operation experimental protocol with sheep red blood cells in irradiated recipients. This kind of experiment is still unsatisfactory technically, it can only be run for a short period of time, and even at the early times the finding of two 'like' cells together can always be rendered meaningless by saying that the two cells can be the progeny of one dividing cell. Nevertheless, bearing in mind these limitations, I still failed to find in spleen sections

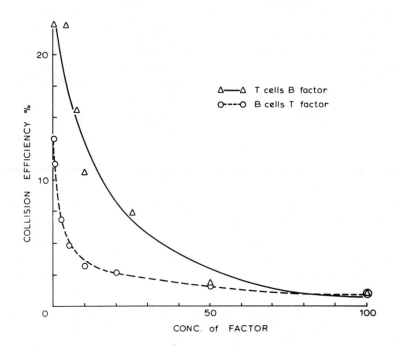

Fig. 3.10 Dose response curves for T and B factors acting on the opposite cell type. Ordinate: adhesiveness (collision efficiency); abscissa, concentration of factor as a percentage of that in the system from which the factor was prepared. Data from Curtis and de Sousa, (1973).

or imprints, any indication that two unlike cells met frequently, the finding of two ^{3}H-labelled thymus or two ^{14}C-labelled marrow cells, on the other hand, was quite common (Fig. 3.11a, 3.11b), suggesting to me that a 'third party' cell is necessary for co-operation *in vivo*.

(4) *Being reunited by the dilution of unlike cells.* It was not until recently that we made the observation that frequent direct T—B cell interactions

Fig. 3.11 Autoradiographs of imprints of spleens removed at 24 h after the intravenous injection of ^{14}C-adenosine labelled bone marrow ^{3}H-adenosine labelled thymus cells and antigen. Note that unlike cell types are found apart (11a), and like cell types frequently seen together (11b).

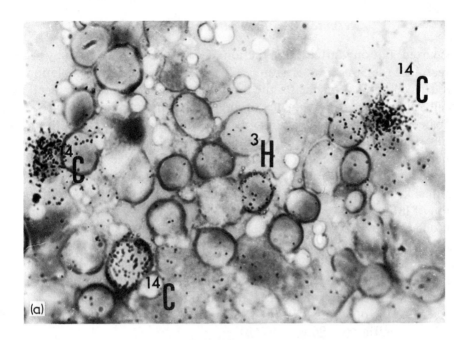

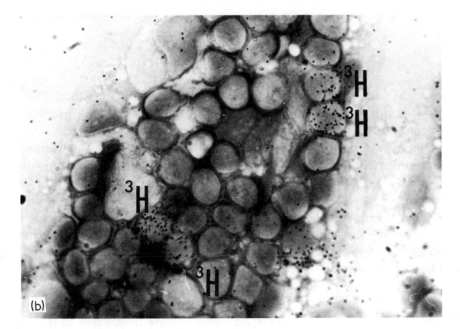

occurred, but only after the induction of partial T cell depletion by ALS treatment (de Sousa and Haston, 1976). It seems that partial depletion of a T cell population creates the opportunity for unlike cells to meet, as well as increasing the frequency of direct B cell interactions. As anticipated, this effect can be counteracted by the administration of thymus cells (Fig. 3.12). We have interpreted these results in the light of Curtis' theory of control of cell positioning (Curtis, 1974) suggesting that the production by T cells of 'interaction—modulation factors' in a normal animal keeps both B cells and T and B cells apart. When the concentration of interaction—modulation factors is reduced by partial reduction of one cell population, as happens after ALS treatment and during the early steps of an immune response in which partial depletion of T cell areas is observed (de Sousa and Parrott, 1967) the chance of direct cell interactions increases. This would constitute a most ingenious device controlling both cell traffic speed and immunological function, of benefit to the operation of any of the currently proposed models of cell co-operation (de Sousa, 1976b).

3.5 TRAFFIC OF 'UNWANTED' CELLS: THE PROBLEM OF METASTASIS

In an adult animal, the physiological rule for red and white blood cells is continuous traffic. Stationary red cells or lymphocytes are generally associated with disease i.e. anaemia and other diseases characterized by lymphocyte infiltration of non-lymphoid tissues, skin, thyroid, liver, etc.. Conversely, the physiological rule of parenchymatous and endocrine organ cells, i.e. skin, thyroid, liver etc, is not to travel beyond short distances of local regeneration processes. Traffic of cells from these organs is generally associated with malignant transformation. Malignant cells leave primary tumour masses, enter the blood or the lymph circuits and are found to settle and proliferate in sites distant from the original tumour. Tumour cells that enter the regional lymph circuit generally metastasize to predictable sites along the draining route, for example, the chain of lymph nodes draining a breast tumour. Tumour cells that enter the blood circuit, however, do not distribute themselves to the sites predicted on the basis of straightforward haemodynamics. Although a great many tumours have metastasis in the lung and the liver, few metastasize to the spleen.

Prostate and thyroid carcinomas have an unusually high incidence of

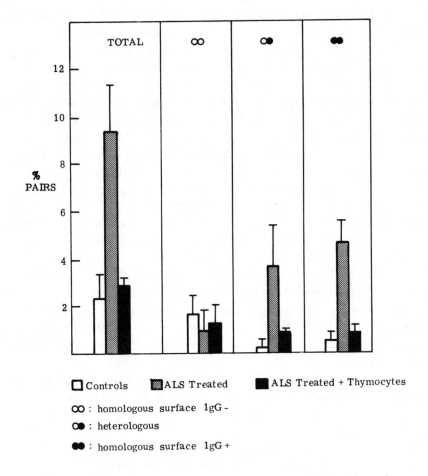

Fig. 3.12 Frequency of direct cell-cell interactions (expressed as % pairs) in lymph node cell suspensions from mice that received one single injection of ALS, 24 h before being killed. Note that the frequency of direct B–B or T–B interactions is much higher in ALS treated mice and corrected by the injection of thymocytes. Data from de Sousa and Haston, 1976.

metastasis in bones, lung carcinomas frequently metastasize to brain, thymic tumours to kidneys, etc.. The earlier literature on this problem is summarized by Willis in a book on *'The spread of tumours in the human body'* (Willis, 1955). From reading Chapter 14 of his book one is led to a paper by Paget who, in 1889 was the first to point out that the distribution of metastasis did not correlate with prediction based on strict knowledge of anatomy of the blood circulation, adding that the problem should be viewed as one of 'seeds' and 'soils', in which the soil was as important to the settling and proliferation of the seed as the seed itself. Reading it one cannot but think of the similarity between Paget's reasoning and Trentin's findings in the work on spleen colonization of bone marrow cells in the mouse (see Section 3.4.1.). The spleen, an organ that clearly has micro-environments inducive to the proliferation of haemopoietic stem cells, has an unusually low incidence of metastases. Early work by Foulds (1932), however, indicates that manipulation of the phagocytic compartment by repeated intravenous injections of trypan blue, increases spleen susceptibility to metastatic growth.

More recently, the question of factors influencing organ specificity of metastatic spread has been approached by Fidler (1973) and Nicolson and Winkelhake (1975). Using mouse B16 melanoma lines variants which acquired enhanced metastatic behaviour to lung after continuous cycling of the tumour cells between *in vivo* lung colonies and *in vitro* cell cultures (Fidler, 1973), Nicolson and Winkelhake (1975) investigated the nature of the adhesive interactions between the tumour variant cells and cell suspensions obtained from either lung, liver, spleen or kidney. The results indicate that the lines selected for more lung metastasis showed highest organ specific aggregation *in vitro* i.e. lung > liver, lung > spleen, lung > kidney.

Thus adhesive interactions between circulating and resident cells seem to play an important role too in the control of metastatic spread.

3.6 THE CENTRIPETAL JOURNEY: UNDERSTANDING HOW AND WHY

I should like to conclude my own journey into this subject by taking the route to the central point where one is forced to ask: how and why?

3.6.1 How does cell traffic occur?

3.6.1.1 Clues from studies of cell traffic during embryonic development
Cell migration in early embryonic development is one of the most extensively studied aspects of cell traffic which has been the subject of numerous reviews in the past (see Trinkaus, 1969; Weston, 1970) and will be reviewed in detail in a forthcoming issue of the present series (Specificity of Embryological Interactions, ed. D. Garrod).

During early embryonic development some cells depart from the primordial structures to migrate and colonize specific environments within the embryo. The two striking examples of this early form of ecotaxis are the migration of neural crest cells and of primordial germ cells. Studies of the fate of neural crest cells derived from homo-specific grafts of neural tube marked with radioisotopes (Weston, 1963; Chibon, 1967) or from heterospecific grafts expressing different biological markers, i.e. exchanging tissues between quail and chick embryo (Le Douarin and Teillet, 1970) have established that, for example, neural cell precursors of pigment cells follow precise routes within or under the ectoderm (Weston, 1963; Chibon, 1967; Le Douarin and Teillet, 1970) to occupy well-defined positions within the epidermis. Other derivatives of the neural crest which also find their precise and separate ways in the developing embryo include neuroblasts, myoblasts and chondroblasts. The development of cartilage seems of particular interest in that it only occurs in the presence of pharynx endoderm and, *in vitro*, chondroblasts seem to develop from those neural crest cells that interact with pharynx endoderm (Epperlein, 1974). The classical studies of the fate of primordial germ cells into the gonadal region (Dubois, 1968) and the more recent analysis of the fate of chick stem cells into quail thymus rudiments (Le Douarin and Jotereau, 1975) further emphasize the precise nature of cell traffic and positioning during the early stages of development.

Perhaps the most relevant aspect of these studies to the present question is that microtechniques have been developed that enable the analysis of cell traffic between tissues, separated by membranes in culture, a technical achievement still to be initiated in the field of lymphoid cell or tumor cell migration. Under culture conditions where all influences of blood supply are eliminated, directional migration of the specific cells to the tissues to be colonized has been observed (Dubois, 1968; Le Douarin and Jotereau, 1975). Other *in vitro* studies of the interaction between neural crest cells and pharynx endoderm, however, failed to demonstrate that directional migration occurs in this system and suggest that contact inhibition is the phenomenon of paramount importance in the positioning of neural crest cells (Epperlein, 1974).

It is conceivable that both chemotaxis and direct cell—cell interactions play causal roles in the early migration and positioning of circulating embryonic cells, depending on the circulating cell's stage of development, surface characteristic and the consequent variable ability to respond to chemotactic stimuli.

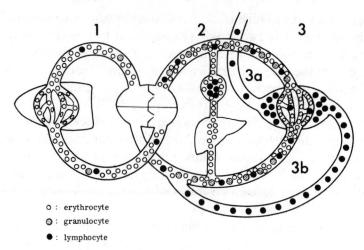

o : erythrocyte
⊚ : granulocyte
● : lymphocyte

LYMPHOCYTE : ENZYMES AND PLANT LECTINS

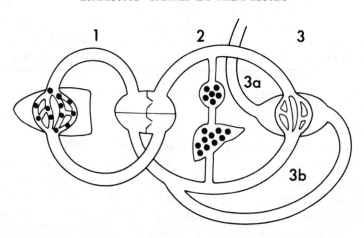

Fig. 3.13 Diagram illustrating the spearate traffic ways of erythrocytes (o),
granulocytes (⊚) and lymphocytes (●) in the blood and lymph circuits
(13a). Modification of red cells generally results in their sequestration in
the spleen and liver (13b).

3.6.1.2 *Traffic in fully developed organisms*

If a fluid-containing tube is connected at one end to a powerful pump,
any particles suspended in the fluid will be forced to travel along the
lumen. When the pump is a heart creating systolic pressures of 120 mmHg
(in man) it is hardly surprising that cells travel fast away from it. The
intriguing questions of cell traffic are posed at those points in the

ERYTHROCYTE: HEAT DAMAGE AND ANTIBODY

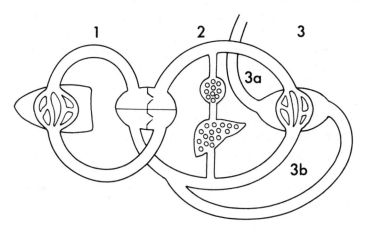

EFFECT OF ANTIGEN

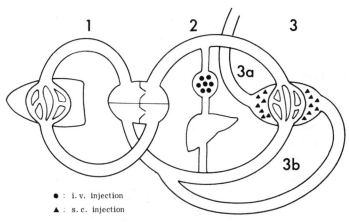

● : i. v. injection
▲ : s. c. injection

Fig. 3.13 *continued* Modification of lymphocytes in their sequestration in spleen, liver or lungs (13c) and administration of antigen (13d) in lymphocyte sequestration in the organ draining the site of injection.

circulation well away from the central pump, with low blood pressures, where different cells types clearly exhibit distinct traffic habits.

In essence, however, the mechanism determining direction, velocity and ultimate destination of traffic of a red cell, a lymphocyte, a mestatasizing tumour cell must be the same i.e. resulting from the direct interaction of the circulating cell with resident endothelial cells or with perivascular cells in the tissues with capillary sinusoids.

Granulocytes seem equipped with the most sensitive detecting

devices, capable of answering any non-specific call from a longer distance than the distance between the interacting surfaces of a red cell and
a macrophage, or two lymphocytes, or a tumour cell and a lung alveolar
cell. One of the early changes occurring in the inflammatory process
takes place in the post-capillary venules and is revealed by the early
change in the rolling of granulocytes along the wall, indicating that
granulocyte traffic too is controlled by direct interactions with resident
endothelial cells at points where chemotactic signals are perhaps received
better. Chemotaxis is most likely an important element in influencing
granulocyte emigration *in vivo* and perhaps one of the reasons why
polymorphs are not found normally in the substance of lymph nodes or
in the white pulp of the spleen is because large collections of lymphocytes do not contain factors chemotactic for granulocytes. In the absence
of lymphocytes e.g. in neonatally thymectomized mice, granulocytes
are seen occasionally in the empty thymus-dependent areas of the lymph
nodes.

In summary (Fig. 3.13) all cells once in the blood circulation are
pushed by the powerful heart pump into a series of parallel capillary
beds from which erythrocytes and granulocytes return via the venous
system. Lymphocytes, on the other hand, emigrate from blood to lymph
in the peripheral lymphoid organs and utilize the lymph circuit to return
to the heart via the thoracic veins. Tumour cells that drain into lymph
nodes often share with lymphocytes the 'exclusive' use of the lymph
route to reach the heart, but they can also enter the blood circulation
from primary tumours which invade arterial or venous organ networks.

Cell traffic velocity through a capillary network is determined by the
nature of the adhesive interactions between the cell in transit and the
endothelial cells lining its route, or, in the organs with capillary sinusoids,
between the cell in transit and surrounding macrophages. Within the
peripheral lymphoid organs lymphocyte traffic velocity is most probably
influenced by the adhesive interactions between T and B lymphocytes.
The exact molecular basis of most adhesive interactions is unknown,
with the exception of the adhesive interaction between macrophages
and antibody or complement coated red cells. Cell traffic velocity is
also impaired by the administration of antigens by a mechanism which,
in essence, must be similar to the one mentioned above, i.e. change in
adhesive interactions between circulating cells and cells in the organ
draining the site of antigen inoculation. Major microvascular changes
take place in this case and they too must be taken into account when
considering changes in cell traffic velocity.

In a few instances, lymphocytes stay in the blood and fail to enter the lymph nodes, whether the answer to the nature of the much publicized interaction between lymphocyte and lymph node post-capillary venules resides in these observations or they are simply an additional example of change in adhesiveness, reduction in this case, remains to be demonstrated.

3.6.2 But why do cells travel?

The reasons for red cell and granulocyte traffic are straightforward compared with the incomprehension that still surrounds lymphocyte traffic. During the red cells' short stay in the capillary alveolar network (nearly 1 s) oxygen and carbon dioxide are exchanged so rapidly that blood leaving the capillaries is virtually in equilibrium with alveolar air. The molecular basis of this process is known: haemoglobin (Hb) is a compound of globin and haem (containing Fe); one molecule of haemoglobin combines with one molecule of O_2 for every atom of Fe it contains, forming oxyhaemoglobin (Hb O_2). The number of red cells per cubic millimeter of blood is 5×10^6. This seems the number of circulating cells required for a vital function, one that is going on all the time. We are breathing all the time, filtering 170 l of fluid through the renal glomeruli every day (estimated blood flow through the kidney per min is 1200 ml) but not being assaulted by bacteria, or fire, or cold, all the time. Thus, granulocyte traffic can be envisaged as an economic measure taken by evolution. By having traffic of granulocytes and the granulocytes equipped with a sensitive detector system responding to long distance non-specific chemotactic stimuli, it is possible to do with 6000 cells ml^{-1} that which would otherwise need at least 5×10^6 cells.

Erythrocyte and granulocyte function are two clearcut examples of how cell traffic evolved to the benefit and the greater efficiency of the function; in the former, cell traffic seems the obvious way of dealing with the problem of delivery of oxygen between the single site of intake and the multiple sites of consumption. In the latter, the traffic of a small highly specialized population seems an economical way of dealing with a problem that throughout the life time of most higher vertebrates may arise only two or three times a month, not every second.

Lymphocytes are neither many in the blood, like erythrocytes, nor sensitive to non-specific chemotactic stimuli, like granulocytes. In the mammal their numbers in intermediate and central lymph are high, they have complex surface structures which include specific receptors for

Table 3.9 Comparison between arrival times of labelled cells to lymph, after intravenous injection

References	Species	Animal	Peak of arrival to lymph
Ellis and de Sousa, 1974	Fish	Plaice	0.5
Bell and Lafferty, 1972	Birds	Duck	5.9–11
Pearson *et al.*, 1976	Mammals	Foetal lamb	12–18
Frost *et al.*, 1976		Young lamb	18–24
Ford and Gowans, 1969		Adult rat	15–20
Sprent, 1973		Adult mouse	18–24
Frost *et al.*,		Adult sheep	27–36

antigen (Wigzell, 1973), they can affect the adhesion of other cells and they circulate continuously between blood and lymph. A measure of of the degree of blood to lymph circulation in the mammal is well illustrated by the recent outstanding work of Pearson *et al.*, (1976) on the ontogeny of lymphocyte traffic in the foetal lamb. Between 70–150 days post-conception, the concentration of lymphocytes in the lymph increases about five fold (from 2.0×10^{-6} ml to 9.0×10^{-6} ml) and during the same period, lymph flow increases about 20 times from 0.7 ml h^{-1} to 22.4 ml^{-1} Moreover, looking at the times labelled cells injected intravenously take to reach the lymph, in the species studied so far (Table 3.9), indicate that time of blood to lymph crossing is proportional to amount of peripheral lymphoid tissue, and the quantity of lymphocytes in lymph proportional to development of thymus function, (see Good and Gabrielsen, 1964).

The argument for blood to lymph circulation being beneficial for the dissemination of immunological memory (Ford and Gowans, 1969; Ford, 1975) is weakened if one remembers that the blood circuit on its own covers all parts of the body and that species with a paucity of lymphocytes in the lymph (i.e. pig) mount adequate secondary immune responses (Binns and Hall, 1965). Pearson *et al.*, (1976) point out that 'in the foetus the recirculating cells cannot be memory cells nor can their pathway of recirculation be directed by any immunological stimulus. Recirculation is thus a physiological property of immunologically virgin lymphocytes and pathways of recirculation must be determined by inherent properties of the cells themselves or the tissues through which lymphocyte traffic occurs'.

One must remember, however, that tumour cells too have properties

Table 3.10* Serum and lymph concentration of albumin, complement, 7S and 19S
antibody in immunized rats†

	Serum concentration ‡	Lymph concentration (% of serum concentration)
Albumin	28.8 g l^{-1}	60
Complement	99 units	32
7S antibody	622 units	53
19S antibody	18.4 units	25

* Data from Kaartinen *et al.*, 1973.
† Immunized either with NIP or DNP hapten.
‡ Logarithmic mean values.

that enable them to survive in the lymph circuit. The question of the
better survival of tumour cells in lymph than blood (Alexander, 1971)
led Kaartinen *et al.*, (1973) to compare the complement and immunoglobulin levels in the serum and the thoracic duct lymph of immunized
rats. The results are summarized in Table 3.10.

On the whole, they confirm the results of Heremans' group (Nash and
Heremans, 1972; Vaerman *et al.*, 1973) indicating that lymph concentration of proteins transudated from blood to plasma, obeys a rule in which
the degree of tansudation is inversely proportional to molecular weight,
i.e. the larger the molecular weight, the lower the concentration in lymph;
mesenteric lymph versus serum ratios of α-macroglobulin and IgM in the
rat, for example, are 0.267 and 0.387 respectively, compared with the
ratios of 0.688 and 0.587 for albumin and transferrin (Vaerman *et al.*,
1973). There are two novel findings in Kaartinen's work, of relevance
to the present discussion: (a) the fact that lymph contains only 32% of
the serum concentration of complement and (b) in a few animals,
simultaneous low levels of antibody and complement were found in the
lymph. Potentially cytotoxic conditions which could influence survival
of cells carrying antigen on the surface, are therefore markedly different
in lymph and serum.

Additional differences between blood and lymph may exist that are
inimical to lymphocyte survival in the blood. The evidence points to
the fact that once a cell has made the blood to lymph crossing, its long
term survival is assured (Caffrey *et al.*, 1962; Rieke and Schwartz, 1967;
Howard, 1972; Sprent and Basten, 1973; Simpson and Cantor, 1975;
Sprent and Miller, 1976).

I am concluding this review by putting forward the proposition that the finding of increasing numbers of long-lived lymphocytes in mammalian lymph is a reflection of:

(1) Stepping up in evolution of specialized thymus function.

(2) A physiological process of natural selection.

Cells that perhaps for their non-adhesive properties and capacity to reduce the adhesion of other cells (Curtis and de Sousa, 1973, 1975) can make the briefest journey in the systemic blood circuit, cross post-capillary venules, enter the lymph circuit, escape death and thus survive 'to remember'.

REFERENCES

Alexander, P. (1971), Factors contributing to the 'success' of antigenic tumours, *Adv. Exp. Biol. Med.*, **12**, 567.

Atherton, A. and Born, G.V.R. (1972), Quantitative investigations of the adhesiveness of circulating polymorphonuclear leukocytes to blood vessel walls, *J. Physiol.*, **232**, 447.

Atherton, A. and Born, G.V.R. (1973), Relationship between the velocity of rolling granulocytes and that of the blood flow in venules, *J. Physiol.*, **233**, 157.

Atkinson, J.P., Schreiber, A.D. and Frank, M.M. (1973), Effects of corticosteroids and splenectomy on the immune clearance and destruction of erythrocytes, *J. Clin. Invest.*, **52**, 1509.

Baez, S. and Kochen, J. (1965), Laser induced microagglutination in isolated vascular model systems, *Ann. N.Y. Acad. Sci.*, **122**, 738.

Bell, E.R. and Shand, F.L. (1975), Changes in lymphocyte recirculation and liberation of the adoptive memory response from cellular recognition in irradiated recipients, *Eur. J. Immunol.*, **5**, 4.

Bell, R.G. and Lafferty, K.J. (1972), The flow and cellular characteristics of cervical lymph from unanaesthetized ducks, *Aust. J. Exp. Biol. Med.* **50**, 611.

Berman, H.J. and Fuhnor, R.L. (1966), in *Blood Flow and Microcirculation* (ed. Charm, S.E. and Kurland, G.S.) (1974), John Wiley and Sons, New York.

Berney, S.N. and Gesner, B.M. (1970), The circulatory behaviour of normal and enzyme altered thymocytes in rats, *Immunology*, **19**, 681.

Binns, R.M. and Hall, J.G. (1965), The paucity of lymphocytes in the lymph of unanaesthetized pigs, *Brit. J. Exp. Pathol.*, **47**, 275.

Bloch, E.H. (1962), A quantitative study of haemodynamics in the living microvascular systems, *Am. J. Anat.*, **110**, 125.

Brown, D.L. (1973), The immune interaction between red cells and leukocytes and the pathogenesis of spherocytosis, Annotation, *Brit. J. Haematol.*, **25**, 691.

Brown, D.L. and Lachman, P.J. (1971), The temporary sequestration of C3-coated red cells (E C43) in the RES of the rabbit: a mechanism for the non-lytic damage of red cells by complement, in *The Reticulo-endothelial System and Immune Phenomena,* (Ed. N.R. Di Luzio), p. 111, Plenum Press, New York.

Brown, D.L., Lachman, P.J.and Dacie, J.V. (1970), The *in vivo* behaviour of complement-coated red cells: studies in C6-deficient, C3 depleted and normal rabbits, *Clin. Exp. Immunol.,* 7, 401.

Burnet, F.M. (1957), A modification of Jerne's theory of antibody production using the concept of clonal selection, *Aust. J. Science,* 20, 67.

Caffrey, R.W., Rieke, W.O. and Everett, N.B. (1962), Radioautographic studies of small lymphocytes in the thoracic duct lymph of rats, *Acta Haematol.,* 28, 145.

Cahill, R.N.P., Frost, H. and Trnka, Z. (1976), The effects of antigen on the migration of recirculating lymphocytes through single lymph nodes, *J. Exp. Med.,* 143, 870.

Cantor, H. and Boyse, E.A. (1975a), Characterisation of subclasses of T lymphocytes at different stages of thymus-dependent differentiation, in *Biological Activity of Thymus Hormones* (ed. van Bekkum, D.W.), p. 78, Kooyker Scientific Publ., Rotterdam.

Cantor, H. and Boyse, E.A. (1975b), Functional subclasses of T lymphocytes bearing different Ly antigens. I. The generation of functionally distinct T-cell subclasses is a differentiative process independent of antigen. *J. Exp. Med.,* 141, 1376.

Celada, F. (1966), Quantitative studies of the adoptive immunological memory in mice: I. An age-dependent barrier to syngeneic transplantation, *J. Exp. Med.,* 124, 1.

Charm, S.E. and Kurland, G.S. (1974), *Blood Flow and Microcirculation,* John Wiley and Sons, New York.

Chibon, P. (1968), Marquage nucléaire par la thymidine tritiée des dérives de la crête neurale chez l'Amphibien urodèle Pleurodels Waltlii Michah., *J. Embroyol. Exp. Morph.,* 18, 343.

Chien, S., Usami, S., Dellenback, R.J. and Gregerson, M.I. (1970), Shear dependent deformation of erythrocytes in rheology of human blood, *Am. J. Physiol.,* 219, 136.

Cooper, M.D., Peterson, R.B.A., South, M.A. and Good, R.A. (1966), The function of the thymus system and the bursa system in the chicken, *J. Exp. Med.,* 123, 75.

Crome, P. and Mollison, P.L. (1964), Splenic destruction of Rh-sensitized and heated red cells, *Brit. J. Haemat.,* 10, 137.

Curtis, A.S.G. (1973), Cell adhesion, *Prog. Biophys. Mol. Biol.,* 27, 317.

Curtis, A.S.G. (1974), The specific control of cell positioning, *Arch. Biol.,* 85, 105.

Curtis, A.S.G. and de Sousa, M. (1973), Factors influencing the adhesion of lymphoid cells, *Nature New Biol.,* 244, 45.

Curtis, A.S.G. and de Sousa, M. (1975), Lymphocyte interactions and positioning. I. Adhesive interactions, *Cell Immunol.,* **19**, 282.

Dalmasso, A.P., Martinez, C. and Good, R.A. (1962), Further studies of suppression of the homograft reaction by thymectomy in the mouse, *Proc. Soc. Exp. Biol. Med.,* **111**, 143.

Davies, A.J.S., Leuchars, E., Wallis, V. and Koller, P.C. (1966), The mitotic response of thymus-derived cells to antigenic stimulus, *Transplantation,* **4**, 438.

Davies, A.J.S., Leuchars, E., Wallis, V., Marchant, R. and Elliot, E.V. (1967), The failure of thymus-derived cells to make antibody, *Transplantation* **5**, 222.

Dintenfass, L. (1964), Rheology of packed red blood cells containing haemoglobins A-A-S-A and S-S, *J. Lab. Clin. Med.,* **64**, 594.

Dresser, D.W.A., Taub, R.N. and Kramtz, A.R. (1970), The effect of localized injection of adjuvant material on the draining lymph node, *Immunology,* **18**, 663.

Dresser, D.W.A. (1961), A study of the adoptive secondary response to a protein antigen in mice, *Proc. Roy. Soc.,* London, (B) **154**, 398.

Dubois, R. (1968), La colonisation des ébauches gonadiques par les cellules germinales de l'embryon de Poulet, en culture, *in vitro, J. Embryol. Exp. Morph.,* **20**, 189.

Durkin, H.G. and Thorbecke, G.J. (1973), Homing of B lymphocytes to follicles: specific retention of immunologically committed cells, *Adv. Exp. Med. Biol.,* **29**, 63.

Durkin, H.G., Caporale, L. and Thorbecke, G.J. (1975), Migratory patterns of B lymphocytes. I. Fate of cells from central and peripheral lymphoid organs in the rabbit and its selective alterations by anti-immunoglobulin, *Cell Immunol.,* **16**, 285.

Einstein, A. (1906), A new consideration of molecular dimensions (in German), *Ann. Phys.* (Leipzig), **18**, 289.

Ellis, A.E. and de Sousa, M. (1974), Phylogeny of the lymphoid system: study of the fate of circulating lymphocytes in plaice, *Eur. J. Immunol.,* **4**, 338.

Epperlein, H.H. (1974), The ectomesenchymal-endodermal interaction system (EEIS) of *Triturus alpestris* in tissue culture. I. Observations on attachment, migration and differentiation of neural crest cells, *Differentiation,* **2**, 151.

Erslev, A.J. and Atwater, J. (1963), Effect of mean corpuscular haemoglobin concentration on viscosity, *J. Lab. Clin. Med.,* **62**, 401.

Fahraeus, R. and Lindquist, R. (1931), Viscosity of blood in narrow capillary tubes, *Am. J. Physiol.,* **96**, 562.

Feldman, M. (1972), Cell interactions in the immune response *in vitro*. II. The requirement for macrophages in lymphoid cell collaboration, *J. Exp. Med.,* **135**, 1049.

Feldman, M. and Basten, A. (1972a), Specific collaboration between T and B lymphocytes across a cell impermeable membrane, *Nature New Biol.,* **237**, 13.

Feldman, M. and Basten, A. (1972b), Cell interactions in the immune response *in vitro*. I. Metabolic activities of T cells in a collaborative antibody response, *Eur. J. Immunol.*, **2**, 213.

Feldman, M., Cone, R. and Marchalonis, J.J. (1973), Cell interactions in the immune response *in vitro*. VI. Mediation by T cell surface monomeric IgM, *Cell Immunol.*, **9**, 1.

Fidler, I.J. (1973), Selection of successive tumour lines for metastasis, *Nature New Biol.*, **242**, 148.

Finch, C.A. (1972), Pathophysiologic aspects of sickle cell anaemia, *Am. J. Med.*, **53**, 1.

Florey, H.W. (1958), *General Pathology* (2nd edn.), Lloyd Luke, London.

Florey, H.W. (1970), *General Pathology* (4th edn.), Llyod Luke, London.

Florey, H.W. and Gowans, J.L. (1958), The reticulo-endothelial system. The omentum lymphatic drainage. The lymphocyte, in *General Pathology* (ed. Florey, H.W.), p. 98, Lloyd Luke, London.

Ford. C.E. (1966), Traffic of lymphoid cells in the body, in *The Thymus*: *Experimental and Clinical Studies*, Ciba Foundation Symposium (ed. G.E.W. Wolstenholme and R. Porter), p. 131, J. and A. Churchill, Ltd., London.

Ford, W.L. (1975), Lymphocyte migration and immune responses, *Progress in Allergy*, **19**, 1.

Ford, W.L. and Gowans, J.L. (1969), The traffic of lymphocytes, *Sem. Haematol.*, **6**, 67.

Foulds, L. (1932), The effect of vital staining on the distribution of the Brown-Pearce rabbit tumour. 10th Scientific Report of the Imperial Cancer Research Fund (London), p. 21.

Freitas, A.A. and de Sousa, M. (1975), Control mechanisms of lymphocyte traffic. Modification of the traffic of ^{51}Cr-labelled mouse lymph node cells by treatment with plant lectins in intact and splenectomised hosts, *Eur. J. Immunol.*, **5**, 831.

Freitas, A.A. and de Sousa, M. (1976a), The role of cell interactions in the control of lymphocyte traffic, *Cell Immunol.*, **22**, 345.

Freitas, A.A. and de Sousa, M. (1976b). Control mechanism of lymphocyte traffic. Altered migration of ^{51}Cr-labelled mouse lymph node cells pretreated *in vitro* with lipopolysaccharide. *Eur. J. Immunol.*, **6**, 269.

Freitas, A.A. and de Sousa, M. (1976c). Control mechanism of lymphocyte traffic. Altered migration of ^{51}Cr-labelled mouse lymph node cells pretreated *in vitro* with phospholipases, *Eur. J. Immunol.*, (in press).

Frost, H. Cahill, R.N.P. and Trnka, Z. (1975). The migration of recirculating autologous and allogeneic lymphocytes through single lymph nodes, *Eur. J. Immunol.*, **5**, 839.

Frost, P. (1974), Further evidence for the role of macrophages in the initiation of lymphocyte trapping, *Immunology*, **27**, 609.

Frost, P. and Lance, E.M. (1974), The cellular origin of the lymphocyte trap, *Immunology*, **26**, 175.

Fulton, G.P., Jackson, R.G. and Lutz, B.R. (1946), Cinephotomicroscopy of normal blood circulation in the cheek pouch of the hamster, *Aietus auratus, Anat. Rec.*, **96**, 537.

Gesner, B.M. (1966), Cell surface sugars as sites of cellular reactions: possible role in physiological processes, *Ann. N.Y. Acad. Sci.*, **129**, 758.

Gesner, B.M. and Ginsburg, V. (1964), Effect of glycosidases on the fate of transfused lymphocytes, *Proc. Nat. Acad. Sci.*, **52**, 750.

Gillette, R.W., McKenzie, G.O. and Swanson, M.H. (1973), Effect of concanavalin A on the homing of labelled T lymphocytes, *J. Immunol.*, **111**, 1902.

Glick, B., Chang, T.S. and Jaap, R.G. (1956), The bursa of Fabricius and antibody production, *Poult. Sci.*, **35**, 224.

Goldstein, G., Scheid, M., Hammerling, U. *et al.*, (1975), Isolation of a polypeptide that has lymphocyte-differentiating properties and is probably represented universally in living cells, *Proc. Nat. Acad. Sci.*, **72**, 11.

Good, R.A. and Fischer, D.W. (1971), (eds.), *Immunobiology*, Sinawer Associates.

Good, R.A. and Gabrielsen, A.E. (1964), (eds.), *The Thymus in Immunobiology*, Harper and Row.

Good, R.A. and Gabrielsen, A.E. (1964), (eds.), *The Thymus in Immunobiology*,

Gowans, J.L. (1957), The effect of the continuous re-infusion of lymph and lymphocytes on the output of lymphocytes from the thoracic duct in unanaesthetized rats, *Brit. J. Exp. Pathol.*, **38**, 67.

Gowans, J.L. (1959), The recirculation of small lymphocytes from blood to lymph in the rat, *J. Physiol.*, **146**, 54.

Gowans, J.L. (1962), The fate of parental strain small lymphocytes in F hybrid rats, *Ann. N.Y. Acad. Sci.*, **99**, 432.

Gowans, J.L. (1965), The role of lymphocytes in the destruction of homografts, *Brit. Med. Bull.*, **21**, 106.

Gowans, J.L. (1966), Life-span, recirculation and transformation of lymphocytes, *Int. Rev. Exp. Path.*, **5**, 1.

Gowans, J.L., and Knight, E.J. (1964), The route of recirculation of lymphocytes in the rat, *Proc. Roy. Soc., B*, **159**, 257.

Grant, L. (1973), The sticking and emigration of white blood cells in inflammation, in *The Inflammatory Process*, (ed. Zweifach, B.W., Grant, L. and McCluskey, R.T.), Voll. II, p. 205, Academic Press, New York.

Greaves, M.F., Owen, J.J.T. and Raff, M.C. (1974), *T and B lymphocytes*, Excerpta Medica, Amsterdam.

Greenblatt, M., Choudari, K.W., Sanders, A.G. and Shubik, P. (1960), Mammalian microcirculation in the living animal: methodologic considerations, *Microvasc. Res.*, **1**, 420.

Griscelli, C., Vasalli, P. and McCluskey, R.T. (1969), The distribution of large dividing lymph node cells in syngeneic recipient rats after intravenous injection, *J. Exp. Med.*, **130**, 1427.

Hall, J.G. (1976), Selective entry of immunoblasts into gut from intestinal lymph., *Nature*, **259**, 308.

Hall, J.G. and Morris, B. (1965), The immediate effect of antigens on the cell output of a lymph node, *Brit. J. Exp. Path.,* **46**, 450.

Ham, T.H., Shen, S.C., Fleming, E.M. and Castle, W.B. (1948), Studies in the destruction of red blood cells. IV. Thermal injury, *Blood.,* **3**, 373.

Hellberg, K., Wayland, H., Rickart, A.L. and Bing, R.J. (1972), Studies on the coronary circulation by direct visualization, *Am. J. Cardiol.,* **24**, 593.

Herman, P.G., Yamamoto, I. and Mellins. H.Z. (1972), Blood microcirculation in the lymph node during the primary immune response, *J. Exp. Med.,* **136**, 697.

Hippocrates, see Littré, 1853.

Hochmutt, R. and Mohandres, N. (1972), Metabolic dependence of red cell shape: observation with scanning electron microscope, *Microvasc. Res.,* **4**, 295.

Howard, J. (1972), The life span and recirculation of marrow derived small lymphocytes from rat thoracic duct, *J. Exp. Med.,* **135**, 185.

Huber, H., Douglas, S.D. and Fudenberg, H.H. (1969), The IgG receptor: an immunological marker for the characterisation of mononuclear cells, *Immunology,* **17**, 7.

Humphrey, J.H. (1955), The mechanism of Arthus reaction. II. The role of polymorphonuclear leucocytes and platelets in reversed passive reactions in the guinea-pig, *Brit. J. Exp. Pathol.,* **36**, 283.

Humphrey, J.H. and White, R.G. (1970), *Immunology for Students of Medicine,* Blackwell, Oxford.

Humphrey, J.H., Parrott, D.M.V. and East, J. (1964), Studies on globulin and antibody production in mice thymectomized at birth, *Immunology,* **7**, 419.

Hurley, J.V. (1963), An electron microscopic study of leucocytic emigration and vascular emigration in rat skin, *Aust. J. Exp. Biol. Med. Sci.,* **41**, 171.

Isles, C. (1974), A study of stem cell/lymphocyte axis. B.Sc. Thesis, University of Glasgow.

Jandl, H.J., Jones, A.R. and Castle, W.B. (1957), The destruction of red cells by antibodies in man. I. Observations on the sequestration and lysis of red cells altered by immune mechanisms, *J. Clin. Invest.,* **36**, 1428.

Jerne, N.K. (1955), The natural selection theory of antibody formation, *Proc. Nat. Acad. Sci.,* **41**, 849.

Kaartinen, M., Kosunen, T.U. and Mäkelä (1973), Complement and immunoglobulin levels in the serum and thoracic duct lymph of the rat, *Eur. J. Immunol.,* **3**, 556.

Kimber, R.J. and Lander, H. (1964), The effect of heat on human red cell morphology, fragility and subsequent survival *in vivo, J. Lab. Clin. Med.,* **64**, 922.

Komuro, K. and Boyse, E.A. (1973), Induction of T lymphocytes from precursor cells *in vitro* by a product of the thymus, *J. Exp. Med.,* **138**, 479.

Kurland, G.S., Charm, S.E. and Tousignant, P. (1968), Comparison of blood flow in a living vessel and in glass tubes, in *Hemorheology* (ed. Copley, A.L.), p. 609.

Lance, E.M. and Taub, R.N. (1969), Segregation of lymphocyte populations through differential migration, *Nature*, **221**, 841.

Lay, W.H. and Nussenzweig, V. (1968), Receptors for complement on leukocytes, *J. Exp. Med.*, **128**, 991.

Le Douarin, N. and Teillet, M-A. (1970), Sur quelques aspects de la migration des cellules neurales chez l'embryon de Poulet étudiée par la méthode des greffes heterospécifiques de tube nerveux, *C.R. Seances Soc. Biol. Fil.*, **164**, 390.

Le Douarin, N. and Jotereau, F.V. (1975), Tracing of cells of the avian thymus through embryonic life in interspecific chimeras, *J. Exp. Med.*, **142**, 17.

Levey, R. and Burleson, R. (1975), The immune conversion of precursor cells by the perfused thymic gland, in *Biological Activity of Thymic Hormones* (ed. van Bekkum, D.W.), p. 103, Kooyker Scientific Publ., Rotterdam.

Littré, E. (1853), Oeuvres Complètes d'Hippocrates, Vol. 8, p. 550, Baillière, Paris, London.

Logan, G. and Wilhelm, D.L. (1963), Ultraviolet injury as an experimental model of the inflammatory process, *Nature*, **198**, 968.

Marchesi, V.T. and Gowans, J.L. (1964), The migration of lymphocytes through the endothelium of venules in lymph nodes: an electron microscopic study, *Proc. Roy. Soc., B*, **159**, 283.

Marsh, G.W., Lewis, S.M. and Szur, L. (1966), Use of ^{51}Cr-labelled heat damaged red cells to study splenic function. I. Evaluation of method. *Brit. Haematol.*, **12**, 161.

Martin, W.J. (1969), Assay for the immunosuppressive capacity of antilymphocyte serum. I. Evidence for opsonization, *J. Immunology*, **103**, 974.

Medawar, P.B. (1958), The Croonian Lecture. The homograft reaction, *Proc. Roy. Soc., B*, **148**, 145.

Merrill, E.W., Benis, A.M., Gilliland, A.R., Sherwood, T.K. and Salzman, E.W. (1965), Pressure flow relations of human blood in hollow fibres at low flow rates, *J. Appl. Physiol.*, **20**, 954.

Micklem, H.S. and Loutit, J.F. (1966), *Tissue Grafting and Radiation*, Academic Press, New York.

Miller, J.F.A.P. (1961), Immunological function of the thymus, *Lancet*, **2**, 748.

Miller, J.F.A.P. (1962), Effect of neonatal thymectomy on the immunological responsiveness of the mouse, *Proc. Roy. Soc.* (Lond.), *B*, **156**,, 415.

Mollison, P.L. (1962), The reticulo-endothelial system and red cell destruction, *Proc. Roy. Soc. Med.*, **55**, 915.

Mongini, P.K.A. and Rosenberg, L.T. (1976), Inhibition of lymphocyte trapping by a passenger virus in immune ascitic tumours: characterisation of lactic dehydrogenase virus (LDV) as the inhibitory component and analysis of the mechanism of inhibition, *J. Exp. Med.*, **143**, 100.

Morell, A.G., Gregoriadis, G., Scheinberg, I.H., Hickman, J. and Ashwell, G. (1971), The role of sialic acid in determining the survival of glycoproteins in the circulation, *J. Biol. Chem.*, **246**, 1461.

Morse, S.I. (1964), Studies on the lymphocytosis induced in mice by *Bordetella pertussis, J. Exp. Med.,* **121**, 49.

Morse, S.I. and Barron, B.A. (1970), Studies on the leukocytosis and lymphocytosis induced by *Bordetella pertussis.* III. The distribution of transfused lymphocytes in pertussis treated and normal mice, *J. Exp. Med.,* **132**, 663.

Morse, S.I. and Bray, K.K. (1969), The occurrence and properties of leukocytosis and lymphocytosis stimulating material in the supernatant fluids of *Bordetella pertussis* cultures, *J. Exp. Med.,* **129**, 523.

Morse, S.I. and Riester, S.K. (1967a), Studies on the leukocytosis and lymphocytosis induced by *Bordetella pertussis.* I. Radioautographic analysis of the circulating cells in mice undergoing pertussis-induced hyperleukocytosis, *J. Exp. Med.,* **125**, 401.

Morse, S.I. and Riester, S.K. (1967b), Studies on the leukocytosis and lymphocytosis induced by *Bordetella pertussis.* II. The effect of pertussis vaccine on the thoracic duct lymph and lymphocytes of mice, *J. Exp. Med.,* **125**, 619.

Muller, A. (1941), *Areh. Kreist. Forsdi.,* **8**, 245. Cited in *Blood Flow and Microcirculation* (ed. Charm, S.E. and Kurland, G.S.), (1974), J. Wiley and Sons.

Munro, A.J. and Taussig, M.J. (1975), Two genes in the major histocompatibility complex control immune response, *Nature,* **256**, 103.

Nash, D.R. and Heremans, J.F. (1972), Intestinal mucosa as a source of serum IgA in the rat, *Immunochemistry,* **9**, 461.

Nicolson, G.L. and Winkelhake, J.L. (1975), Organ specificity of blood borne tumour metastasis by cell adhesion? *Nature,* **255**, 230.

Nims, J.C. and Irwin, J.W. (1973), Chamber techniques to study microvasculature, *Microvasc. Res.,* **5**, 105.

Owen, J.J.T., Cooper, M.D. and Raff, M.C. (1974), *In vitro* generation of B lymphocytes in mouse foetal liver, a mammalian 'bursa equivalent', *Nature* **249**, 361.

Paget, S. (1889), The distribution of secondary growths in cancer of the breast, *Lancet,* i, 571.

Parrott, D.M.V. (1962), Strain variation in mortality and runt disease in mice thymectomised at birth, *Transplant. Bull.,* **29**, 102.

Parrott, D.M.V. (1976), The gut associated lymphoid tissues and gastrointestinal immunity, in *Immunological Aspects of the Gastrointestinal Tract and the Liver* (ed. A. Ferguson and R. McSween), Medical and Technical Publishing Co.

Parrott, D.M.V. and Ferguson, A. (1974), Selective migration of lymphocytes within the mouse small intestine, *Immunology,* **26**, 571.

Parrott, D.M.V. and de Sousa, M. (1971), Thymus-dependent and thymus-independent populations: origin, migratory patterns and life span, *Clin. Exp. Immunol.,* **8**, 663.

Parrott, D.M.V., de Sousa, M.A.B. and East, J. (1966), Thymus-dependent areas in the lymphoid organs of neonatally thymectomised mice, *J. Exp. Med.,* **123**, 191.

Parrott, D.M.V., Rose, M.L., Sless, F., Freitas, A. and Bruce, R. (1975), Factors which determine the accumulation of immunoblasts in gut, in *Future Trends in Inflammation* (ed. Giroud). p. 32 Birkhäusen Verlag, Basel B.

Pearson, L.D., Simpson-Morgan, M.W. and Morris, (1976), Lymphopoiesis and lymphocyte circulation in the sheep fetus, *J. Exp. Med.,* **143**, 167.

Raff, M.C. (1969), Theta isoantigen as a marker of thymus-derived lymphocytes in mice, *Nature,* **224**, 378.

Rieke, W.O. and Schwartz, M.R. (1967), The proliferative and immunologic potential of thoracic duct lymphocytes from normal and thymectomised rats, in *The Lymphocyte in Immunology and Haemopoiesis* (ed. Yoffrey, J.M.), p. 224, Edward Arnold, London.

Rifkind, R.A. (1966), Destruction of injured red cells *in vivo, Am. J. Med.,* **41**, 711.

Roelants, G.E., Loor, F., van Boehmer, H., Sprent, J., Hägg, L.B., Mayor, K.S. and Ryden, A. (1975), Five types of lymphocytes characterised by double immunofluorescence and electrophoretic mobility, *Eur. J. Immunol.,* **5**, 127.

Roitt, I. (1974), *Essential Immunology,* (2nd edn.), Blackwell Scientific Publications.

Rose, M., Parrott, D.M.V. and Bruce, R. (1976), Migration of lymphoblasts to the small intestine. I. Effect of *T. spiralis* on the migration of mesenteric lymphoblasts and mesenteric T lymphoblasts. *Immunology,* (in press).

Scheid, M.P., Hoffman, M.R., Komuro, K., Hämmerling, U., Abbott, J., Boyse, E.A., Cohen, G.H., Hooper, J.A., Schulof, R.S. and Goldstein, A.L. (1973), Differentiation of T cells induced by preparations from thymus and by non-thymic agents, *J. Exp. Med.,* **138**, 1027.

Schlesinger, M. (1970), Anti-theta antibodies for detecting thymus-dependent lymphocytes in the immune response of mice to SRBC, *Nature,* **226**, 1254.

Schlesinger, M. and Israel, E. (1974), The effect of lectins on the migration of lymphocytes *in vivo, Cell Immunol.,* **14**, 66.

Schlitt, L.E. and Keitel, L.G. (1960), Renal manifestation of sickle cell disease: a review, *Am. J. Med. Sci.,* **239**, 773.

Schreiber, A.D. and Frank, M.M. (1972a), The role of antibody and complement in the immune clearance and destruction of erythrocytes. I. *In vivo* effects of IgG and IgM complement-fixing sites, *J. Clin. Invest.,* **51**, 575.

Schreiber, A.D. and Frank, M.M. (1972b), Role of antibody and complement in the immune clearance and destruction of erythrocytes. II. Molecular nature of IgG and IgM complement-fixing sites and effect of their interaction with serum, *J. Clin. Invest.,* **51**, 583.

Shand, J. and de Sousa, M. (1974), The simultaneous detection of ^{3}H and ^{14}C labelled cells by double-layer auto-radioagraphy, *J. Immunol. Methods,* **6**, 141.

Simonsen, M. (1962), Graft-versus-host reactions. Their natural history and applicability as tools of research, *Progr. Allergy,* **6**, 349.

Simpson, E. and Cantor, H. (1975), Regulation of the immune response by subclasses of T lymphocytes. II. The effect of adult thymectomy upon humoral and cellular responses in mice, *Eur. J. Immunol.,* **5**, 337.

Singh, U. (1975), Adenyl cyclase activation and expression of theta-antigen on
  foetal thymocytes, in *The Biological Activity of Thymic Hormones* (ed.
  van Bekkum, D.W.), p. 29, Kooyker Scientific Publ., Rotterdam.
de Sousa, M. (1971), Kinetics of the distribution of thymus and bone marrow cells
  in the peripheral organs of the mouse: ecotaxis, *Clin. Exp. Immunol.,* **9**,
  371.
de Sousa, M. (1973), The ecology of thymus-dependency, in *Contemporary Topics
  in Immunobiology* (ed. Davies, A.J.S. and Carter, R.L.), Vol. 2, p. 119,
  Plenum Press, New York.
de Sousa, M. (1976a), Microenvironment to a lymphoid cell is nothing more than
  interaction with its neighbours, *Adv. Exp. Biol. Med.* **66**, 165.
de Sousa, M. (1976b), Ecotaxis, ecotaxopathy and lymphoid malignancy. *Immuno-
  pathology of Lymphomas,* eds. R.A. Good and J. Twomey, Plenum Press,
  New York.
de Sousa, M. and Haston, W. (1976), Modulation of B cell interactions by T. cells,
  Nature **260**, 429.
de Sousa, M. and Parrott, D.M.V. (1967), Definition of a germinal center area as
  distinct from the thymus-dependent area in the lymphoid tissue of the
  mouse, in *Germinal Centres in The Immune Response* (ed. Cottier, H. and
  Odartchenko,), p. 361.
de Sousa, M.A.B. and Parrott, D.M.V. (1969), Induction and recall in contact
  sensitivity, *J. Exp. Med.,* **130**, 671.
de Sousa, M., Ferguson, A. and Parrott, D.M.V. (1973), Ecotaxis of B cells in the
  mouse, *Adv. Exp. Med. Biol.,* **29**, 55.
Sprent, J. (1974), Migration and life span of circulating B lymphocytes of nude
  (nu/nu) mice, in Proc. First. Internat. Workshop on nude mice (ed.
  Rygaard, J. and Povlsen, C.O.), p. 11, Fisher Verlag, Stuggart.
Sprent, J. (1976), Fate of H-2 activated T lymphocytes in syngeneic hosts:
  I. Fate in lymphoid tissues and intestines traced with 3H-thimidine
  ^{125}I-Deoxyuridine and [51]chromium, *Cell Immunol.,* **21**, 278.
Sprent, J. (1976b), Recirculating lymphocytes, in *The Lymphocyte Structure
  and Function* (ed. Marchalonis, J.J.), Marcel Dekker Inc., New York.
Sprent, J. and Miller, J.F.A.P. (1976), Fate of H-2 activated T lymphocytes in
  syngeneic hosts: III. Differentiation into long-lived recirculating memory
  cells, *Cell Immunol.,* **21**, 314.
Sprent, J. and Basten, A. (1973), Circulating T and B lymphocytes in the mouse.
  II. Life span, *Cell Immunol.,* **7**, 40.
Strober, S. (1972), Initiation of antibody responses by different classes of lymphocytes.
  V. Fundamental changes in the physiological characteristics of virgin
  thymus-independent (B) lymphocytes and 'B' memory cells, *J. Exp. Med.,*
  136, 85.
Strober, S. and Dilley, J. (1973), Biological characteristics of T and B memory
  lymphocytes in the rat, *J. Exp. Med.,* **137**, 1275.
Taub, R.N. (1974), Effects of concanavalin A on the migration of radioactively
  labelled lymphoid cells, *Cell Immunol.,* **12**, 263.

Taub, R.N., Rosett, W., Adler, A. and Morse, S.I. (1972), Distribution of labelled lymph node cells in mice during the lymphocytosis induced by *Bordetella pertussis, J. Exp. Med.,* **136**, 1581.

Till, J.E. and McCulloch, E.A. (1961), A direct measurement of the radiation sensitivity of normal mouse bone marrow cells, *Radiat. Res.,* **14**, 213.

Tillich, G., Mendoza, L., Wayland, H. and Bing, R.J. (1971), Studies of the coronary microcirculation of the cat, *Am. J. Cardiol.,* **27**, 93.

Tillmanns, H., Ikeda, S., Hansen, H. , Sarnea, J.S.M., Faurel, J.M. and Bing, R.J. (1974), Microcirculation in the ventricle of the dog and turtle, *Circulation Res.,* **34**, 561.

Trentin, J.J., McCarry, M.P., Jenkins, U.K., Gallagher, M.T., Speirs, R.S. and Wolf, N.S. (1971), Role of inductive microenvironments on hemopoietic (and lymphoid?) differentiation and role of thymic cells in eosinophilic granulocyte response to antigen, *Adv. Exp. Med. Biol.,* **12**, 289.

Trinkaus, J.P. (1969) *Cells into organs.* Prentice Hall, Englewood Cliffs.

Uzsoy, N.K. (1964), Cardiovascular findings in patients with sickle cell anaemia, *Am. J. Cardiol.,* **13**, 320.

Vaerman, J.P., Andre, C., Bazin, H. and Heremans, J.F. (1973), Mesenteric lymph as a major source of IgA in guinea-pigs and rats, *Eur. J. Immunol.,* **3**, 580.

van Bekkum, D.W. (1975), *The Biological Activity of Thymic Hormones,* Kooyker Scientific Publ., Rotterdam.

Virchow, R. (1858), Lecture IX, Pyemia and leucoytosis, in *Cellular Pathology* (1860), Churchill, London.

Waksman, B.H., Arnason, B.G. and Jancovic, B.B. (1962), Role of the thymus in immune reactions in rats. III. Changes in the lymphoid organs of thymectomized rats, *J. Exp. Med.,* **116**, 187.

Warner, N.L. and Szenberg, A. (1962), Effect of neonatal thymectomy on the immune response in the chicken, *Nature,* **196**, 784.

Warner, N.L., Szenberg, A. and Burnet, F.M. (1962), The immunological role of different lymphoid organs in the chicken. I. Dissociation of immunological responsiveness, *Aust. J. Exp. Biol. Med. Sci.,* **40**, 373.

Weston, J.A. (1970), The migration and differentiation of neural crest cells, *Adv. Morphol.,* **8**, 41.

Whitaker, S.R.F. and Winton, F.R. (1933), The apparent viscosity of blood flowing in the isolated hind limb of the dog and its variation with corpuscular concentration, *J. Physiol.,* **78**, 339.

Wiederman, M.P. (1962), Lengths and diameters of peripheral arterial vessels in the living animal, *Circ. Res.,* **10**, 886.

Wiederman, M.P. (1963), Dimensions of blood vessels from distributing artery to collecting vein, *Circ. Res.,* **12**, 375.

Wigzell, H. (1973), The functional significance of surface structures on lymphocytes, *Ann. Immunol.* (Inst. Pasteur), **124c**, 7.

Wilkinson, P.C. (1976), The adhesion, locomotion and chemotaxis of leucocytes, in *Inflammation and Anti-inflammatory Drugs* (ed. Ferreira, B.H. and Vane, J.R.), Springer Verlag.

Williams, R.M., Chanana, A.D., Cronkite, E.P. and Waksman, B.H. (1971), Antigenic markers on cells leaving the calf thymus by way of the efferent lymph and venous blood, *J. Immunol.*, **106**, 1143.

Willis, R.A. (1955), *The Spread of Tumours in the Human Body*, Butterworths Medical Publications.

Wolf, N.S. and Trentin, J.J. (1968), Haemopoietic colony studies. V. Effect of haemopoietic organ stroma on differentiation of pluripotent stem cells, *J. Exp. Med.*, **127**, 205.

Woodruff, J. (1974), Role of lymphocyte surface determinants in lymph node homing, *Cell Immunol.*, **13**, 378.

Woodruff, J. and Gesner, B.M. (1968), Lymphocytes: circulation altered by trypsin, *Science*, **161**, 176.

Woodruff, J. and Gesner, B.M. (1969), The effect of neuraminidase on the fate of transfused lymphocytes, *J. Exp. Med.*, **129**, 551.

Zatz, M.M. and Lance, E.M. (1971), The distribution of ^{51}Cr labelled lymphocytes in antigen stimulated mice. Lymphocyte trapping, *J. Exp. Med.*, **134**, 224.

Zatz, M.M., Goldstein, A.L., Blumenfeld, O.O. and White, A. (1972), Regulation of normal and leukaemic lymphocyte and recirculation by sodium periodate oxidation and sodium borohydride reduction, *Nature New Biol.*, **240**, 252.

Zweifach, B.W. and Intaglietta, M. (1968), Mechanism of fluid movement across single capillaries in the rabbit, *Microvas. Res.*, **1**, 83.

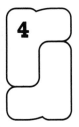

Incompatibility in Flowering Plants

D. LEWIS F.R.S.
Department of Botany and Microbiology,
University College London,
Gower Street,
London.

Receptors and Recognition, Series A, Volume 2
Edited by P. Cuatrecasas and M.F. Greaves
Published in 1976 by Chapman and Hall, 11 New Fetter Lane, London EC4P 4EE
© Chapman and Hall

4.1 INTRODUCTION

Plants have no immunological system producing antibodies or capable of graft rejection and other parameters of cell-mediated immunity [1]. Successful grafting between species and even genera is common and is a well known practical means of propagation. The few unsuccessful grafts are caused by mechanical or nutritional difficulties. No evidence for the presence of immunoglobulins or a histoincompatibility system, encoded by genes, has been found. But plants have specialized recognition systems which are as precise and as characteristic of the individual species as the major histoincompatibility systems in mammalian animals: their function is to increase the efficiency of sexual reproduction.

In contrast to the immune systems of vertebrate animals and possibly invertebrates also (see Chapter 2 by F.M. Burnet in Volume One of this series) the plants recognition and reproduction system is not programmed to recognize and respond to non-self but rather to recognize and discriminate against self. This gives non-self, or non-genetically identical pollen, a selective advantage in fertilization.

A typical example of the scope of one system can be found in a single field of red clover which has been found to contain an estimate of 211 different alleles controlling self- and cross-incompatibility [2, 3]. Every plant is self-incompatible but the finding of a pair of plants which are cross-incompatible would be expected once in 22 245 plants tested.

The cellular components of recognition are the male pollen and the female style. The style has a stigmatic surface for the adherence and germination of pollen and a conducting tissue between the stigma and the ovary. This is usually solid tissue or in some monocotyledons it is a hollow cylinder lined by special stigmoid cells. Pollen grains have an almost infinite variety of shapes, sizes and fine structure of the surface and sub-surface, making possible the identification of genera and sometimes species from living or fossil pollen alone. The exposed stigmatic cells of the style also vary but to a lesser extent than pollen. Supplementing and sometimes reinforcing this morphological variation is a more subtle molecular variation which is often correlated with the morphological differences but with an independent causation. It is generally believed that this morphological and molecular variation of pollen and style in plants has evolved to meet two main requirements of sexual reproduction. One is *selective pollination* to provide efficient transfer of pollen of the right species with transporting agents as haphazard as wind and some indiscriminating insects. This is achieved mainly by complementary sizes and surface sculpturing, and to a lesser degree by the spatial positioning of pollen and stigma. The other is selective *fertilization* to provide an efficient outbreeding system. Approximately 95% of the flowering plants are hermaphrodite with male and female organs in close proximity. Outbreeding and random mating is achieved mainly by reducing or preventing self-fertilization; in some systems sibfertilization, which is of less significance, is also reduced, without restricting random cross fertilization. This is achieved by a selective contraceptive system without loss of fertility of pollen or eggs.

The flowering plants comprise some tens of thousands of species in more than 200 families and it is not surprising that in this large group there are several basic types of pollen and style recognition systems. In some species the selective pollination system is loosely correlated with the selective fertilization, as in the sporophytically controlled* incompatibility in the cabbage family; or they are closely correlated as in heterostyled species, such as *Primula*; or there is no correlation, as in the gametophytic† incompatibility of clover, cherries and other plants. The prevention of self-fertilization is, in some systems, at the point of penetration of the stigmatic outer membrane or it may be at varying distances in the conducting tissue of the style or even in the ovary. *The reaction is an oppositional one between* products of identical alleles; in all systems it is a surface interface phenomenon and, in several species, a protein is the basis *of the inhibition.* The structure and origin of the outer layers of the pollen grain are of prime importance in the surface relations of pollen and stigmas. This has been well reviewed by Heslop Harrison [4]. The outer layer, exine, has a complex chemical nature which is highly resistant to acetolysis and gives the

---

* *Sporophytic control* refers to pollen characters which are not controlled by the genes in the haploid nucleus but by genes in the diploid mother plants which produce the pollen.

† *Gametophytic control.* Pollen grains being gametes with a haploid nucleus have some of their characters, including incompatibility in some systems, controlled by the gene or genes in the haploid nucleus.

pollen grain its indestructible properties as a fossil. Its importance in the present context is its sculpturing which in some species has been shown to complement the sculpturing on the stigmatic surface. However, there is no indication that it has highly specific variation which would account for the specificity of self-incompatibility. The origin of the exine has been in dispute but the best evidence is that it has been derived from the pollen mother cell. The material often present on, and in the cavities of, the exine is lipidic in nature and is produced in the sporophytic tapetum. Although much is known about the genetical control of the systems, there are many unsolved basic problems. The origin and evolution of the systems is not clear, the origin of the large number of alleles, whether by mutation or recombination, has no support from extensive experiments and, finally, the precise molecular and biochemical basis is obscure. The active incompatibility proteins diffuse rapidly out of pollen grains into isotonic media which is more than a pointer to their role in pollen allergy.

4.2 THE SYSTEMS

The incompatibility systems in flowering plants have been classified mainly on morphological and genetical criteria. Morphologically there are two basic types: *heteromorphic* and *homomorphic*. The names are self-explanatory; the heteromorphic incompatibility shows gross morphological differences of style and pollen between the inter fertile mating types, the homomorphic systems show no morphological differences. The heteromorphic systems have developed a certain degree of efficiency of cross pollination by elaborate positioning of pollen and stigma and by pollen and stigmatic cell size and sculpturing. There is naturally a limit to the effective subdivision of such gross morphology and this has restricted the number of compatible mating types, based upon molecular differences, to two in distylic species and three in tristylic species. The homomorphic systems are not restricted by morphology and have developed more efficient outbreeding by increasing the number of compatible mating types to tens of thousands and leaving the pollination to the mercy of indiscriminate wind and insects. This has been achieved by a large series of multiple alleles of one, two or three genes. There are also major differences in the genetical control of the pollen which are correlated with differences in pollen cytology. Every variation of sexual differentiation within a hermaphrodite kingdom has been exploited in the pursuit of outbreeding and heterozygosity.

4.2.1 Heteromorphic incompatibility

Heteromorphic incompatibility has been found in 23 families and 134 genera [5]. These families are diverse, spread across the plant kingdom and appear to have a polyphyletic origin, indicating that heteromorphy has many independent origins, although arguments can be advanced for an unified origin [6]. The full syndrome of morphological characters differentiating the two types in dimorphy and three types in trimorphy are:

(1) Pollen shape, size, sculpturing and composition of the outer layers;
(2) Pollen position in the flower as determined by another length or point of another insertion in the corolla tube;
(3) Stigmatic papillae in shape, sculpturing and surface composition;
(4) Length of style and in *Primula spp.* the area of cross section of the conducting tissue in the style;
(5) A molecular incompatible system between pollen and the stigma or pollen tube and the conducting tissue.

Not all heteromorphic species have the full complex of characters, but the sea lavender, *Limonium vulgare* [14], has most and is illustrated in Fig. 4.1. As an illustration of the complementary fit in the sculpturing of pollen and stigma in the compatible mating, the scanning E.M. photograph of *Limonium meyeri* [7] is a good example (Fig. 4.2). The five main character pairs are combined together to make an inter-compatible pair of mating types. Long style with large stigmatic cells and short anthers with small pollen is found in the long styled plant and short style with small stigmatic cells and high anthers with large pollen in the short styled plant. In addition is the incompatible—compatible reaction which in some genera is the inhibition of pollen germination on the stigma and in others the inhibition of the pollen tubes in their growth through the style. Plants with atypical combinations of these five characters, called homostyles, are found either as a rarity in many populations [8] or in high proportion in some rare populations in *Primula vulgaris* [9] or as an entire but rare population in *Pemphis acidula* [10]. The most extensive study of homostyles in *Primula spp* made by Ernst has revealed uncoupling between style type and anther type and between pollen size and anther height. The incompatibility reaction and pollen size has not been uncoupled in any homostyle described. The whole complex, as will be shown later, is under the control of a tightly linked

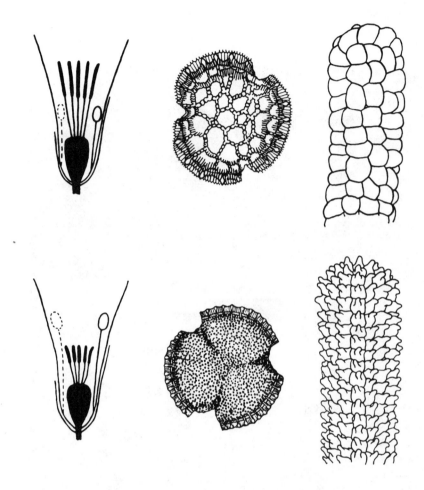

Fig. 4.1 Left: diagrammatic section through flower; Centre: pollen grain; Right: stigma; Upper row: Long styled flower; Lower row: short styled flower, of the distylic species, *Limonium vulgare*. By kind permission of Professor Herbert Baker and the Editor of *Evolution*.

group of five or more genes. The failure to find a separation of pollen size from incompatibility indicates the pollen stigma surface interaction or incompatibility as the prime beginning of the whole complex.

The efficiency of the characters varies greatly between species and genera, both for pollination and for incompatibility. The few studies on natural pollen loads distributed by insects shows that correct pollination is high (96% and 88%) in *Limonium meyeri* [7]. In contrast, Levin [11],

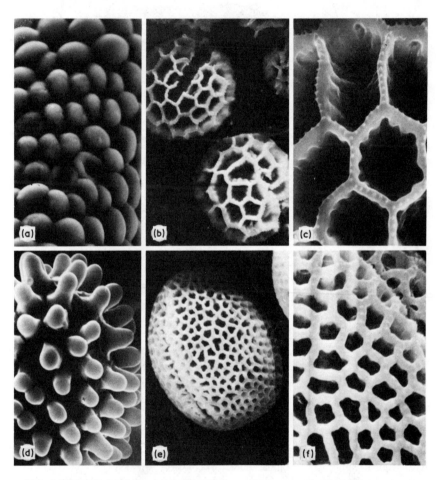

Fig. 4.2 Detailed structure of the pollen and stigma surface in the two types of
the distylic species, *Limonium meyeri.* By kind permission of Dr. R. Dulberger
and the Royal Society. (a) Cob stigma. (x 800), (b) Type A pollen grains.
(x 800), (c) Exime of A pollen. (x 2400), (d) Papillate stigma. (x 800),
(e) Type B pollen grains, (f) Exime of B pollen. (x 2400).

found 0.8% on long styles and 10% on short styles in *Lithospermum
carolinensis.* From the limited evidence, the relative position of anthers
and stigmas does not appear to contribute greatly to compatible pollination. The pollen and stigma surfaces are of greater importance. The

waxless pollen and stigmas of the long-styled *Linum grandiflorum* reduce adherence of the self pollen to zero so that self pollen grains do not adhere, swell or germinate [12]. A similar non-adhesion is found in other *Linum spp.* [13]. Further evidence that pollen dimorphism is the prime initiator among the morphological characters is found in a comparative study of species and genera in the Plumbaginaceae. Baker [14], has traced the development of dimorphism from monomorphic species through simple dimorphic pollen, dimorphic pollen and stigma and finally heterostyly. All these dimorphic and monomorphic species are self-incompatible which indicates the sequence: incompatibility, pollen dimorphism, heterostyly, thus incompatibility predates dimorphism.

The surface properties of the pollen grain, which determine specific adherence and incompatibility where it is a germination effect, are located in the exine and sexine [15]. The tryphine, which coats the pollen and is lipoprotein, is synthesized in the maternal tissues of the tapetum and deposited on the pollen grains [16, 17]. This is in keeping with the genetic control of the pollen characters of dimorphism and incompatibility. It is the genotype of the sporophyte and not the gametophyte which is in control.

Trimorphism, with three types of pollen, stigmata, style lengths and anther heights, has been found in only three families [5, 18]. These three families all have flowers with two distinct whorls of anthers, one opposite the sepals and the other opposite the petals. In each mating type there are two levels of anthers in the same flower, corresponding to the stigma levels in the other two mating types. It would appear that the restriction of trimorphism to these families is due to the necessity of two differentiated whorls of anthers for the development of two levels of anthers [10]. The exception that supports this rule is found in *Narcissus triandrus* which has a spurious trimorphism in which the incompatibility is not related to the trimorphism [19]. *Narcissus* is typical of its family by not having two differentiated whorls of anthers, all the anthers being symmetrically inserted on the petaloid perianth. It would appear that to cause the complete switch within the same flower of anther height, pollen size and incompatible specificity, it is necessary to have a basic differentiation in tissue areas.

In both dimorphic and trimorphic species the pollen size is closely correlated with the length of the compatible style and with the rate of pollen tube growth, so that the time from pollination to the fusion of egg and pollen nuclei, despite a three-fold difference in style length, is constant. This implies a delicately timed signalling and transmission system between stigma and ovule in order to prepare the egg cell for fertilization. Finally, an important correlation between pollen cytology and the incompatibility system has been found by Brewbaker [20]. Most families of flowering plants have either binucleate or trinucleate pollen grains. The binucleate pollen grains have a delayed division of the generative nucleus in the pollen tube. Similarly, families show a basic difference in the cytokinesis at meiosis [21]. The four products of meiosis can be divided either simultaneously at the end of meiosis or successively after the first and second division. The full correlation between these two characters and the incompatibility system will be made later but suffice it to point out here that all four types are found in heteromorphic species, but that binucleate pollen grains are found in species where pollen tube growth is inhibited and trinucleate where pollen germination is inhibited.

Long homostyle forms are found in tristylic *Lythrum salicaria* and distylic *Fagopyrum esculentum*. The self compatible homostyles in the dimorphic *Primula spp.* are also predominately long styled; the short homostyle is rare. However, in *Limonium spp.* and *Gelsemium elegans* the self compatible homostyles are exclusively short styled. Another consistent difference between long and short styled plants in *Primula* is the greater fertility of short styled plants when compatibly pollinated [18, 22, 23]. The full significance of the different homostyles and fertility is not clear but it is possibly related to the genetic dominance of one type.

4.2.2 Homomorphic incompatibility

Homomorphic incompatibility is more common and widespread than heteromorphy. It is found in about half the families of flowering plants, and some families such as the Solanaceae and Rosaceae have a majority of self incompatible species, and those that are self compatible have become so as a secondary development. It appears to be the most efficient random outbreeding mechanism that any group of organisms has developed. As its name implies, there are no morphological features of the plant, including the pollen, that distinguish the mating types. It is because of this lack of morphological differences that the system has exploited multiallelism and attained greater efficiency.

There are two basic systems, gametophytic and sporophytic. The one-gene gametophytic system with its gametophytic control of the pollen reaction is found exclusively in families with binucleate pollen. The gametophytic system with two genes is found in families with trinucleate pollen and successive cytokinesis. The sporophytic system, with sporophytic control of the pollen, is found exclusively in families with trinucleate pollen and simultaneous cytokinesis [20, 21]. The full significance of these correlations with the pollen cytology is not clear but as suggested by Pandey a reasonable explanation could be that the early determination of sporophytic control is also reflected in an early development of the nuclei and a late development of the dividing cell walls. The two systems also differ in their place of reaction between pollen and style. In all gametophytic systems known the pollen on an incompatible style germinates, penetrates the stigma surface and later becomes inhibited at some point in the conducting tissue. This can vary in different species from a few mm inside the style, as in *Oenothera organensis,* to near the base of the style, as in *Solanum spp.* In some, the tips of the incompatible pollen tubes are swollen, some are burst and in *Tradescantia virginiana,* the author (unpublished) has seen the pollen tubes grow in the stigmatic cells to the surface of the conducting tissue, turn round and grow part of the way in the opposite direction; an observation which, if understood, may help in the interpretation of surface interaction and directed growth.

An E.M. study of pollen tube growth in compatible and incompatible styles by Dickinson and Lawson [24] in *Oenothera organensis* has shown that the growth of the pollen tube wall, after the first elongation, is effected by the addition of fibrils derived from non-cellulosic material in dictysome vesicles which are associated with mitochondria. Both compatible and incompatible pollen tubes are indistinguishable in this process of wall formation and the main difference found is that incompatible tubes have a greatly reduced level of carbohydrates, which are probably the raw materials for the wall fibrils. Perhaps the most significant finding in the study is that, although the final inhibition of the pollen is in the conducting tissue, there are striking differences in carbohydrate mobilization in the pollen grain within minutes of contact with the stigma.

The pollen style reaction in the sporophytic system is exclusively at the stigmatic surface. Pollen germination is inhibited to 90—95% and those grains that do germinate send out a short tube which does not penetrate the wall of the stigmatic cell [25, 26]. This incompatible tube stimulates the production of callose in the stigmatic cell centred on the point of contact [17, 27]. If the stigmatic cells are artificially removed, incompatible pollen will germinate and grow normally down the style and achieve fertilization, showing that the reaction is strictly localized in the stigmatic papillae.

4.3 GENETICS

4.3.1 Heteromorphic

The genetical control of incompatibility in plants is one of its most intriguing features. Although the fine structure and mapping of the controlling genes are not known as they are for certain genes in bacteria, the major properties of incompatibility genes are known in many species. The details of genetic evidence up to 1954 has been reviewed [28] and this review will include a summary of the earlier systems and a more detailed account of some newly discovered systems with more than one controlling supergene. It is necessary to make clear the meaning of terms used. From the tests of segregation and complementation the **S** and **Z** factors controlling incompatibility are single segregating heredity units and in this sense can be called genes. However, in the heteromorphic system the number of disparate characters encoded and the occurrence of rare recombinants show that the **S** factor is a complex of some five or six genes tightly linked together. The homomorphic system is also controlled by a tightly linked complex of genes as revealed by mutation. These complexes may be regulated by control elements as in the *lac* complex in *Escherichia coli* and could be termed an operon. Without evidence for such regulatory elements, the incompatibility complexes should be referred to as a supergene, which has no control connotations and is 'a group of genes acting as a mechanical unit in particular allelomorphic combinations' [29].

The two mating types of dimorphic species, long styled, pin, and short styled, thrum, differ in that one is homozygous for the recessive supergene and the other is heterozygous. In *Primula, Linum* and *Fagopyrum spp.* the long styled form is the homozygote and the short styled is the heterozygote. In the Limoniaceae, Baker [30] has found evidence for heterozygosity of the long styled form. The first suggestion that the complex of characters was controlled by a supergene was by Ernst in 1936 [8], who postulated three genes **G A P**; **G** for gynecium, **A** for anther, **P** for pollen. This was later elaborated [28] to include the stigmatic cells **S** and the incompatibility reactions of pollen and style **Ip, Is**. From the frequency of the occurrence of different homostyles the most probable map order is given below.

	Short Styled					Long Styled				
G	S	Is	Ip	P	A	g	s	is	ip	p a
g	s	is	ip	p	a	g	s	is	ip	p a

The genetic control of the pollen size and incompatibility is from the mother plant, the sporophyte. This is well illustrated by the pollen of a short styled plant, half of which carries S and half s. All the pollen has the size and incompatibility reaction characteristic of S. The S superallele is active in the pollen mother cell and/or the pollen nutritive tissue, the tapetum, before segregation of the alleles at meiosis. The complete dominance of the S over s in both pollen mother cells and the style and the sporophytic control of the pollen are absolutely essential for the successful functioning of this outbreeding system.

The genetic control of trimorphism is by two supergenes S and M. In the three families containing trimorphic species, Lythraceae, Oxalidaceae and Pontederiaceae, the long styled is a double recessive **ss mm**, the mid styled is **ss Mm** or **ss MM** and the short styled is **Ss Mm** or **Ss mm**. The S and M alleles are dominant and S is epistatic to M which makes the order of dominance and epistacy as short-mid-long [31−33]. The only exception to this dominance order is found in a species of *Oxalis* in which the M gene is epistatic to S [34, 35]. The pollen control is sporophytic and again the dominance and sporophytic control are essential for the proper functioning of this recognition system.

One important aspect of incompatibility, apart from its efficiency as an outbreeding system, is the ease with which it can be disrupted by a rare event, and if temporally or locally advantageous, revert to a viable inbreeding population. The different incompatibility mechanisms have different causes of breakdown which reflect their genetics. The heteromorphic systems are disrupted by a rare cross-over in which the components of the supergene are rearranged to produce fully viable, fertile and self-compatible homostyles. Many homostyle species of *Primula* have been derived in this way from heterostyled ancestors. Polyploidy, which is a common feature of plants, has no effect on the expression of S and s and in consequence heterostyled species are frequently polyploid. It will be apparent why polyploidy has no effect on the expression of heterostyly after consideration of the gametophytic system. Polyploidy, however, does have an indirect effect on heterostyly. It appears to relieve the tight linkage of the supergene, possibly by the shift of chiasmata in a multivalent. Homostyle plants have been found after artificial polyploidy in *Primula obconica* [23] and *Lythrum salicaria* [36]. Heteromorphic incompatibility has also been weakened by the selection of the polygenic background [22]. A summary of the genetics of heteromorphism is given in Table 4.1. The striking feature to note in contrast to the other systems is the restricted number of alleles

Table 4.1 The incompatibility systems with their genetic features and the method of breakdown.

Systems		No of Supergenes	No of alleles	Allele interactions		Breakdown	
				Pollen	Style	Method	Product
Heteromorphic							
Distyly		1	2	Dominance	Dominance	Recombination	SC and SI Homostyles
Tristyly		2 IC	2 + 2				
Homomorphic	Gametophytic	2 D	Many	Not applicable	Codominance	Polyploidy	SC
		2 IC				Mutation	
		3 IC					
		1					
Homomorphic	Sporophytic	2 IC	Many	Dominance and codominance	Dominance and codominance	Mutation	SC
		3 IC					

IC = incompatible complementary, in which two or more genes complement to give the incompatible reaction;

D = duplicative, in which two genes have the same duplicate action, either one of which if allelically matched will give the incompatible reaction;

SC = self-compatible; SI = self-incompatible.

enforced by the shortcoming of floral morphology as a multivariate and yet discrete parameter.

4.3.2 Homomorphic

The two basic homomorphic systems are distinguished by the period and place of S gene action both in relation to the pollen and the style. The encoding of the pollen by the S gene, whether gametophytic or sporophytic, has the most important consequences because, as will be seen, it determines the dominance and codominance requirements of the multiallelic series which is present in both systems. Examples of other morphological and biochemical characters of pollen grains which have no connection with incompatibility are known to have either sporophytic or gametophytic control. The round and long pollen of *Lythrus odoratus* first described by Bateson *et al.*, in 1905 [37] is sporophytically controlled with long dominant to round, while the waxy and starchy pollen of *Zea mays* was found 20 years later by Demerec [38] and Brink [39] to be gametophytically controlled by the haploid pollen genotype, where dominance cannot be expressed. When these and other pollen characters are classified on their genetic encoding, it is found that the gametophytically encoded characters (eight known) are always *internal,* such as the type of carbohydrate synthesized within the pollen grain, whereas the sporophytically encoded characters (of which 12 are known) are mostly *external*, such as the shape of the pollen grain, which is a property of the wall, the surface sculpturing of the exine, the surface colour, or the covering of tryphine. Rarely are they internal, such as a deficiency of carbohydrate.

The developmental stage of the S gene action in gametophytic incompatibility is known precisely from the expression of mutations induced by a pulse acting mutagen such as X-irradiation [40]. If irradiation is given to prepollen mother cells or pollen mother cells up to the second division of meiosis, the mutations induced are expressed in the resulting pollen grains. If the irradiation is given at the first signs of the tetrad formation, no mutations are expressed in the resulting pollen. The timing of S gene action in the sporophytic system cannot be so easily determined because the recessive nature of most mutations precludes their observance, even if the mutational event occurred before the gene product had been formed. With regard to the tissue of S gene action, the implication of tryphine, which is synthesized in the tapetum, as the source of the incompatibility protein is in keeping with sporophytic encoding.

4.3.3 Gametophytic system

The early work on gametophytic systems revealed a similar genetic
control in a wide variety of species of the Dicotyledons. It was a single
segregating unit **S** which had multiple alleles. Later the unit was found
to be not a single gene but a single supergene of several tightly linked
genes. In several species linkage of the supergene to other genes controlling, for example, flower colour was found. Lundqvist [41] and Heyman
[42] found gametophytic incompatibility under the control of two
unlinked supergenes **S** and **Z** in the grasses. The grasses are Monocotyledons
which are taxonomically considered to be primitive. Lundqvist has also
found incompatibility under the control of three separate supergenes in
a primitive Dicotyledonous family, the Ranunculaceae. He has argued
that the three gene system is primitive and that the one gene system is
derived and more advanced [43]. From other considerations and the
finding of a three gene sporophytic system (see next section), the author
is in favour of this view and considers that, if this can be substantiated,
it puts incompatibility systems and their origin in an entirely new light.

The gametophytic system with one **S** supergene, in all species examined
and with all alleles tested, exhibits independent action of the pairs of
alleles in the style which is diploid somatic tissue. One allele is never
dominant so that all alleles show codominance. In the haploid pollen
grain dominance cannot be expressed so that each pollen grain is expressing the **S** reaction specific to the allele it carries. The recognition system
of incompatibility is *self recognition* to produce an incompatible rejection.
It is an *oppositional* action between products of the same allele to produce a positive inhibition of pollen tube growth. For the system to
achieve a successful block to self fertilization the codominance in the
style is essential with gametophytic control of the pollen, because all
plants are heterozygous for **S** alleles and there will be two different
pollen reactions for each plant, which must be matched by the same two
reactions in the style. It is, therefore, not surprising that in all species
spread over some ten or more disparate families this is the rule. A common origin of self-incompatibility in these families would explain the
uniformity but the view favoured by the author is that the system has
evolved separately in different families but has converged to the same
basic end point because of the rigid requirements of the system. Any
possible allele interactions in the haploid pollen are irrelevant to the
system and, therefore, would not be rigorously selected.

Such potential interactions, however, irrelevant as they are in nature,
have been examined and have given important information bearing upon

the nature of the gene products and the nature of dominance in the sporophytic system. The first indication of a remarkable effect of S allele interaction in pollen came from a tetraploid sport of a pear variety, Fertility [44] and a colchicine-induced tetraploid in *Petunia* [45]. The diploid is self incompatible, the autotetraploid sport is self compatible. This effect has since been found, to varying extents, in five other species including ones from other families. The effect was found to be caused solely by the production of diploid pollen grains with two different S alleles. The presence of four alleles in the tetraploid style or two copies of the same allele in the pollen did not affect the normal recognition reaction [46]. It was as if the two different allele products in the pollen could not recognise the same allele products in the style. This was explained at the time as competition, or in more modern terms negative complementation, between the alleles. The explanation at the protein product level was that the two alleles competed by aggregating to form a hybrid product which would be ineffective in recognising the two components in the style. This might be by the hybrid aggregate acquiring a new specificity or by losing all specificity. The basic explanation has been substantiated but it has not been possible to decide between the two precise alternatives. In fact, it is probable that, depending upon the pair of alleles, both alternatives would be realized because not all allele pairs show this negative complementation. Some, and probably the majority, show one allele to be dominant over another in the pollen grain. This breakdown of the system by polyploidy gives an insight into the exploitation by the system of all aspects of gene and gene product interactions. The fact that haploid pollen grains are not subject to allele interactions, which would disrupt the system, leaves certain gene products, such as multimeric proteins, available for exploitation by the system. The strict codominance maintained in the style is carried so far that in a tetraploid style of *Oenothera organensis* with four different S alleles, all four show codominance despite the fact that the same alleles in the pollen show either dominance or negative complementation. This points strongly to the S allele product in the style being restricted to a monomer, or a dimer with one S allele polypeptide and a second unrelated polypeptide as a neutral cofactor. In the pollen it is assumed that multimers of the S polypeptide are regularly formed and in normal haploid pollen these must be homomultimers.

The complex structure of the S supergene was revealed by mutation studies [47–50]. By using the self style as a highly efficient sieve for mutant S alleles, self-compatible plants were obtained. Mutants, changed in their pollen activity alone or in both pollen and style activity, showed that the **S** supergene was composed of at least three separate genes. A typical mutation would completely obliterate the unique specific allele action in the pollen, but the same unique specificity in the style is unimpaired. Brewbaker and Natarjan [50] made the important discovery in *Petunia* that self-compatible mutant plants had a normal diploid complement of chromosomes and a fragment. This was readily explained if it was assumed that the fragment carried an **S** allele so that some pollen grains carrying the fragment would have two different **S** alleles and hence would be self-compatible by competition, as in a tetraploid. Pandey [51] found in *Nicotiana spp* that all self-compatible mutants which had been produced by ionizing radiations contained a fragment, whereas the majority which were obtained spontaneously without treatment did not carry a fragment. This suggested that the fragment has not only the function of producing self-compatibility by competition but also acts as a complementary cover for a probable recessive lethal effect of the induced **S** mutation.

From complementation studies there is some evidence that the pollen deficient **S** mutants in *Oenothera organensis* are in a regulator or complementary gene and not in the structural gene for the **S** protein [52]. The evidence for this came from testing the interaction between an unmutated S_2 allele and a mutant S_4' in a diploid pollen grain. S_2 and S_4 in a diploid pollen grain are codominant so that the diploid pollen grain expresses both S_2 and S_4 reaction. The mutant S_4' in a haploid pollen grain has lost its S_4 reaction but in the style, *even when not accompanied by a different allele,* has full S_4 reaction. A diploid pollen grain with $S_2 S_4'$ has both S_2 and S_4 reaction which shows that S_2 allele has complemented S_4' to allow its specific S_4 expression in the pollen. Clearly the mutation has affected a gene of the complex which is necessary for some factor common to all **S** alleles and which is essential for the reaction in the pollen but not in the style.

One of the problems of the **S** system is the origin of new alleles. From extensive mutation studies, involving the testing of 10^9 or more pollen grains no fully operative new alleles were produced either by mutation or recombination at meiosis, although the generation of new alleles at some stage in the evolution of the system would have to be at a frequency not less than 1 in 10^{-4} [53, 54]. Recently there are reports of new alleles after enforced inbreeding in *Trifolium* [55], *Lycopersicon* [56] and *Nicotiana* [57]. In *Lycopersicon* the mutation at its first occurrence appears to affect only the style, and after further inbreeding the new specificity is expressed in both pollen and style. These types of mutations would not be selected by the technique of the stylar sieve of self pollination used in most S mutation studies. De Nettancourt and Pandey favour a recombinational origin of the new alleles and that recombination in the S supergene is strictly controlled, but this control is relaxed by inbreeding. These promising pointers to the origin of new specificities only emphasize the complete absence of any hard evidence which leaves the question as open as it is for histocompatibility [58] and γ globulin genes in mammals or the genes for ciliate antigens in ciliates.

The discussion has been confined to species in the most advanced group of plants, the Dicotyledons. In a more primitive group, the Monocotyledons, the grasses are considered to be in a primitive position. A two gene self-incompatibility system has been found in several species of grass including the cultivated rye. The two genes are unlinked and designated S and Z [41, 42] and indicated in Table 4.1 as Homomorphic. Gametophytic 2 IC. The system is of great interest because there are multiple alleles at each gene and the two genes complement to produce an incompatible reaction. The general rule is that an SxZy pollen is inhibited only in a style that contains both Sx and Zy; it is not inhibited in a style containing Sx and any allele of Z except Zy or in a style containing Zy and any allele of S except Sx.

In view of the great specificity and multiplicity of the reaction, the two complementary (2 IC) gene system must mean that either:
(1) There is one specific reaction which is determined by a molecule with at least two active sites, one encoded by the S gene and the other by the Z gene; or
(2) There is a pathway of two or more steps in series which ends in the incompatibility reaction, and two of these steps are mediated by a specific recognition of S products in one step and a Z product in the second step.

The author [59] has favoured the first explanation which leads to the hypothesis that the active molecule is a multimer and probably a dimer in which a polypeptide is contributed from S and Z. Lundqvist found that, contrary to one gene gametophytic systems, tetraploidy had no effect on the expression of S and Z alleles in diploid pollen. Dominance and competition were not found. This can be readily explained if we assume that dimers or multimers can only form between an S and a Z allele product. The complementary nature of the S and Z products would preclude dimerization between two similar products of S and Z. In the one gene system it is presumed that a multimer is always formed in the pollen between two **S** products and in the haploid pollen grain there would always be homomultimers, whereas in the diploid pollen grain there would be no molecular basis for preventing **S** heteromultimers forming. An interesting and unique exception of the one gene gametophytic system which does not show dominance and competion in diploid pollen is found in *Tradescantia paludosa,* another Monocotyledon. This can be contained within the general scheme by Lundqvist's suggestion that such a lack of interactions in diploid pollen is a basic necessity for the evolution of the two gene system in grasses, which to be successful should not and does not show epistatic interactions between the genes. The author favours the view that *Tradescantia* had two genes in its past evolution but one has become homozygous and, therefore, genetically silent, but still participates in the basic incompatibility reaction.

An additional and important complication will arise if the three gene system suggested for *Ranunculus acris* and *Beta vulgaris* [43] can be substantiated. It could swing the evidence away from the dimer hypothesis to a multi-step pathway, for it would necessitate the assumption of a molecule with three cooperating active sites each of which has multiple specificities, or three steps in a series pathway, each of which has a specific and multiple recognition. A full discussion of the one, two and three gene systems will be made after the sporophytic system has been reviewed.

Lundqvist made the reasonable suggestion that the **S** and **Z** genes arose by duplication and were followed by some subsequent differentiation to give the complementary action between them. This duplication should not be confused with the duplicate genes found in *Physalis ixocarpa.* In this case the two genes are undifferentiated duplicates at the gene and phenotype level. A matching of an allele of either gene in pollen and style is enough to give the incompatibility reaction. The consequences of these two kinds of multi gene action on the efficiency of the systems as outbreeding mechanisms is quite different and again of sufficient importance to be discussed later.

4.3.4 Sporophytic system

One of the most unusual genetic features about the sporophytic homomorphic system is that the large multiple allelic **S** gene shows a complexity of dominance and codominance in allelic pairs, which can be arranged in a series of interlocking hierarchical groups and less commonly in a cyclical group. It was found almost impossible to fit such dominance relations into a system of dominance which depended on quantitative differences of concentration of gene products and critical thresholds. It is, however, conceivable in terms of multimeric molecules which have not only the necessary wide variation in the covering and uncovering of active sites but also contain an exponential system for multiplying the number of heteromultimers at the expense of homo-multimers. The power of this exponential system depends upon the number of units in the multimer. It, therefore, makes an enlarged scale from which critical thresholds can be selected.

The first example of the system was found by Correns in 1912 [60] in the Cruciferous plant, *Cardamine pratensis.* The data supported sporophytic control of the pollen, four alleles, two pairs with codomin-ance, four pairs with dominance, expressed similarly in pollen and style. In the 1950s the system was found in other Cruciferous species and in the Compositae [61—64]. Many more alleles were identified and pairs of alleles which showed different dominance relations in pollen and style were found. The most extensive work on one species by Thompson [65] on the cultivated crops of *Brassica oleracea* has revealed at least 40 different alleles which show the full range of dominance, codomin-ance and competition relations. It must be remembered that the labour involved in finding a new allele increases exponentially with the total number known, so it can be concluded that the potential number is much higher than 40. The proportion of pairs showing dominance is 47%, codominance 51% and competition 2%. These different types of dominance are found about equally in pollen and style. This should be contrasted with the gametophytic system, Table 4.1.

For the system to work as a block to self-fertilization, the sporophytic control and dominance in pollen and style are both essential. Unlike the gametophytic system, there is no breakdown with polyploidy. This is in full agreement with the basic hypothesis because the dominance relations in the diploid sporophytic tissue would all be selected for the correct interactions of two different alleles both for pollen and style.

Within the sporophytic system there is a delicate balance between dominance and codominance. If all allele pairs are codominant a species cannot survive with less than four alleles because, with less, all polli-nations would be incompatible. Even to obtain 50% of random pollinations compatible would require 8 alleles. With dominance the system would work with 2 alleles, which gives 50% of compatible pollination. On the other hand dominance gives less protection against cross fertilization between sibs than does codominance. It would appear that a mixture of the two kinds of dominance, which is found, is the optimum both for efficiency of the breeding system and for the molecular interactions of the gene products. If the multimer hypothesis can be substantiated, the sporophytic system would require a multimer in the style as well as in the pollen.

The sporophytic system offers considerable technical difficulties in the interpretation of breeding results because of the great number of different patterns of compatibility relations within a diallel cross of progeny from two parents. In the gametophytic system with two different patterns in a one gene system and six different patterns in a two gene system, the precise fitting of genotypes to breeding groups is simple. With the sporophytic system, each pair of alleles can have one of four different dominance relations; the number of different patterns with two genes has an upper limit of 4^8. Clearly an exact fit is impracticable. By using two simple parameters, the percentages of reciprocally incompatible and non reciprocally incompatible crosses, a discriminative analysis can be made. This has been applied to data obtained in *Eruca sativa*, a Cruciferous plant [66]. A three gene system is the minimum to explain the results. Each of the three genes has to be allelically matched to produce the incompatibility reaction. An application of the incompatible parameter method to the data on the gametophytic system in *Ranunculus acris* and *Beta vulgaris* gives a perfect fit to the results, which gives further support to the three gene interpretation.

4.3.5 Sporophytic–Gametophytic system

To complete, but not as yet to clarify, the picture, an exceptional type of incompatibility is present in *Theobroma cacao*. Self-incompatible pollen tubes grow at a normal compatible rate down the style, the pollen tube releases its nucleus into the embryo-sac and either fails to fertilize the egg or fertilizes and subsequently aborts after a few cell divisions of the endosperm [67]. Breeding studies have revealed a genetic control with at least five alleles, and sporophytic control of the ovules. The evidence for the type of pollen control is not conclusive [68] but the most reasonable explanation is based upon gametophytic control [69]. If this can be substantiated, *Theobroma* will be unique not only in the position and mode of the block to fertilization but also in having a genetic control which is sporophytic on the female side and gametophytic on the male side.

4.3.6 Efficiency and evolution of incompatibility

It has been generally agreed, with Darwin, Darlington and Mather, that self-incompatibility from a teleological point of view has one main function, that is to prevent self-fertilization and to encourage random mating in order to maintain a large pool of genetic heterozygosity, which is considered essential for any long term evolution and adaptation to occur. Random mating implies cross mating between individuals of the population including siblings at a rate proportional to the frequency of the individuals. The importance of reducing sib-mating has not been generally agreed. Continued sib-mating, like selfing, is a well known effective but slower method of inbreeding. Unfortunately there is no direct estimate available of the amount of pollination between sibs. However, from studies of pollen dispersion over distance, it has been estimated that only 1% of the pollen is dispersed at a distance of 20 m, whether transported by wind or insects [70]. We are even more ignorant about the dispersal of seeds, and again what evidence there is points to a limited distance in most terrestrial plants. Another variable to consider, particularly in static organisms, is the density of the population. Allowing for all these factors, the maximum sib-pollination would occur with sparsely placed individuals with poor seed dispersal. With this special pleading the author is taking the view that the frequency of sib-pollination in terrestrial plants is significantly above that expected from random pollination within a population. Other factors to be considered are the effectiveness of the pollen in the population as a whole and whether ineffective, i.e. incompatible, pollen can block fertilization with simultaneous or later compatible pollen, as it can in *Theobroma*. This last point is of major importance and all incompatible systems, except that in *Theobroma*, have avoided this. The importance of cross compatibility of the pollen in the population as a whole is not easy to assess, but judging from the devices to ensure effective pollination such as large stigma, attractive devices for insects, and the sculpturing and covering of pollen, this is of considerable significance and would suggest, for example, the proportion of 50% of ineffective pollen in a distylic system is wasteful.

A property of self-incompatibility that is important in evolution is the ease with which it can, by a rare mutation or recombination, breakdown and turn, without complete disruption, to selfing. This is something that cannot be achieved in an outbreeding system based upon separation of the sexes in different individuals as in animals. Such events are rare enough not to be a burden on the outbreeding population but can be rapidly exploited if selfing becomes either a necessity or a short term advantage. Not only are local inbred populations of a species known, such as homostyle populations, in an otherwise distylic species but also whole species that continually self-pollinate are found in a genus or family that is composed mainly of outbreeding species. The evidence in most cases is that the inbreeders have been derived from an outbred ancestor. Examples of such derived inbreeders are found in all types of self-incompatibility, so that all have this potentiality for a viable change to inbreeding. The only minor point of difference between the incompatibility systems is the positive effect of polyploidy in the gametophytic system which results in self-compatibility. This has tended to restrict the gametophytic system to diploid species whereas no such restriction is found in heterostyly and the sporophytic system. All the self-incompatibility systems prevent self-fertilization, which leaves the only variable of importance in the efficiency as the proportion of compatible sib-matings and the proportion of general compatibility in the population. With self-fertilization at zero the efficiency of a system can be measured by the fraction $\frac{\text{Gen. Com.}}{\text{Sib. Com.}}$. The relative efficiencies of the three main incompatibility systems, known for many years, are: distyly 1.0, one gene gametophytic 2.0 and one gene sporophytic 2.0–4.0. The extremely low efficiency of distyly is partly compensated by the morphological features which encourage effective transport of legitimate pollen. This does not affect the efficiency fraction but it does increase both the general and the sib-compatibility.

These considerations would not have been worthy of a restatement if it had not been for the important discovery of the new systems with two and three genes with IC action. Both with the gametophytic and the sporophytic system, this increase in the number of genes drastically reduces the efficiency index by increasing sib-compatibility. In the gametophytic system the values are one gene 0.5, two genes 0.72, three genes 0.9, and in the sporophytic system the corresponding figures are 0.25–0.5, 0.5–0.86, 0.75–0.97. It is clear that the complex two and three gene systems are less efficient than the simple one gene systems, and this fact throws an entirely new light on the origin, the evolution and, possibly, the molecular basis of incompatibility. It makes the evolution of multigenic systems from one genic to be highly improbable. The complex nature of the highly specific complementation between the alleles would be unlikely to arise unless it had very great advantages; the disadvantage of sib-mating must settle the question. On the other hand, the evolution from a multigenic to a one genic system has at least two advantages. It is more efficient and with one gene in control it has the potentiality to be able, by a rare mutation, to turn to self-fertilization. The means by which the change from multi to one gene is accomplished are readily available. Homozygosity of one or two genes is a frequent and a normal feature of some individuals in a multigenic IC system. Such homozygosity could be readily established, particularly by the sib-mating which is more frequent in a multigenic system. Once homozygosis has occurred only one gene would be segregating, the others would be active but genetically silent. If this is the mode of origin of a one gene system, it should be possible by interbreeding plants from distant populations to demonstrate the presence of a second gene, because any one of the two or three genes might have become homozygous in a particular population. The fact that multigenic systems are found in primitive species also supports the primitive status of the systems.

The question to be answered is how did the system originate in such a complex form, if later simplification can make it more efficient? The efficiency of all the systems depends largely on the great number of alleles. If, at the early development of the system, alleles were limited to two or three, there would be a great advantage of multiplying the number of genes. But it is tempting to speculate that the incompatibility system was superimposed upon a multi-step metabolic or katabolic pathway which, by its very nature, demanded several structural and control genes. From the molecular point of view the genetic evidence points either to a complex multimer, in which three different polypeptides contribute, or to the multi step pathway.

4.4 BIOCHEMISTRY AND PHYSIOLOGY OF INCOMPATIBILITY

4.4.1 Protein participation

Serological tests using rabbit antisera and the precipitin ring, and later Ouchterlony plate, techniques showed that pollen of *Oenothera organensis*, with its gametophytic system, contained a protein which was serologically distinct from the four **S** alleles tested [71]. Later tests with Ouchterlony plates and gel electrophoresis confirmed the findings and showed that the serological specific protein was as high as 20–40% of the total pollen protein and that it diffused out of intact pollen grains into isotonic agar medium in 15–30 min [72] (Fig. 4.3). Furthermore, the diffusate from single pollen grains could be identified by the presence

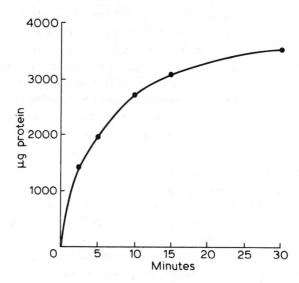

Fig. 4.3 Diffusion of protein including the S protein from intact pollen grains of *Oenothera organensis*, $S_6 S_6$. Note that the diffusion is complete in 30 min. The incompatible reaction as measured by pollen tube growth at the same temperature, 26°C, is complete and irreversible in the same period of time (see p. 193). From Lewis, Burrage and Walls, *J. Exp. Bot.*, 1967.

or absence of precipitin rings in agar containing S-specific antisera [73]. In another species with the gametophytic system, *Petunia hybrida,* pollen and stylar proteins have been identified serologically for three alleles, and the pollen and style S proteins are serologically similar and specific for each allele [74]. A pollen protein is also implicated in a plant with the sporophytic system, *Brassica oleraceae,* using electrophoresis and immunodiffusion with three S alleles [75]. Linskens [76] found by electrophoresis protein fractions which are found only in pollinated styles; one fraction, *Z,* is found only in compatibly pollinated styles and the fractions, *X* and *Y,* are found only in incompatibly pollinated styles. He has also shown that glucose is associated with all three fractions. Radioactive labelling revealed that the glucose comes from the pollen and the protein fraction from the style [77].

It should be stressed that all the serological and electrophoretic observations are only correlations of pollen—stylar reactions and serological or electrophoretic separations. The genetic results, however, demand a molecule with great informational content and with conformational variation as well as cellular activity and, therefore, indicate a

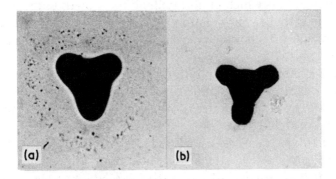

Fig. 4.4 Pollen grains of *Oenothera organensis* (a) S_6 pollen grain (b) S_3 pollen grain; both pollen grains have been on agar with rabbit antiserum produced by injection with S_6 pollen protein. Note the zone of precipitation around (a).

protein. This gives some confidence in the interpretation of serological tests with the mutant S_4' which had lost the pollen S activity but retained in full the S_4 activity in the style ([73] See Section   ). The S_4' pollen contained a serologically distinct fraction which was indistinguishable from the fraction found in the unmutated S_4 [52]. This indicates that the mutation does not affect the specific protein but a common factor, either of low molecular weight or another protein. This is in complete agreement with the complementation studies in diploid pollen where another allele, S_2 restored S_4' activity in an $S_2 S_4'$ pollen grain.

There is no available direct information to support a protein–protein interaction in the heteromorphic system although the indirect evidence supports such a view, at least in some species. The genetics with only two alleles does not demand a protein and in fact the only direct evidence on the physical nature of the primary reaction is in the distylic *Linum grandiflorum,* in which the osmotic pressure of pollen and style are implicated. A ratio of osmotic pressures of pollen/style of 4:1 is found in compatible combinations short x long and long x short. In the incompatible combination short x short, the ratio is 7:1 and the pollen germinates, penetrates a few microns into the style and bursts. In the combination long x long, the ratio is 5:2 and the pollen does not become turgid or germinate [78].

4.4.2 Secondary and side effects

If we define a glycosylated protein–protein interaction as the primary incompatibility trigger, there is no real experimental evidence to define

the secondary effects which inhibit pollen tube growth in the gametophytic system or pollen germination in the sporophytic system. The
secondary or side effects which have been identified in the gametophytic
system are:

(1) an initial higher rate of respiration [76];

(2) changed quantities of free glucose, fructose and sucrose [79, 80];

(3) changed quantities and patterns of free nucleotides, amino acids,
    RNA and protein [81];

(4) a greater frequency of callose plugs in the incompatible tubes [82].

All these indicate a radical effect on carbohydrate utilization and protein
synthesis which are particularly suggestive in relation to the electron
microscope studies in which interference with the utilization of pollen
tube wall fibres has been found [24].

4.4.3 Attempts to show incompatibility *in vitro*

The major difficulty in elucidating the biochemistry of the reaction is
the lack of an *in vitro* test. Many unpublished and some published
attempts have been made but no successful method has been found, at
least for the gametophytic system. Pollen tube growth in culture medium
supplemented with stylar extracts have failed to give results. The use of
paper chromatography with two different stylar extracts and then used
as a pad for growing the pollen of the two forms of the distylic *Primula
obconica,* revealed spots which inhibited pollen tube growth but these
were not related to the incompatibility system. The nearest approach to
a satisfactory method is a semi *in vitro* test which has been obtained in
the sporophytic system where the tryphine of incompatible pollen can
stimulate the production of the wound callose in the living stigmatic
cells, which is a symptom of the incompatibility reaction [27]. The
difficulty found in simulating self-incompatibility *in vitro* emphasizes
the membrane and surface nature of the recognition. A method using
membranes and a staining technique might succeed.

4.4.4 Condition of the plant before the reaction

Almost any condition which reduces the vigour and metabolism of the
plant reduces the strength of the incompatibility reaction. The senescence
occurring at the end of the season has a weakening effect to cause 'end
season pseudofertility'. Keeping the plant in the dark for three days was
first found to weaken incompatibility in *Oenothera organensis* [83] and

a similar effect has been found in *Petunia* [81] and other plants. Special treatments of the style such as soaking in water or weak boric acid solution or water treatment at 50°C [79] or treatment with gaseous carbon dioxide [84], all weaken the incompatibility reaction. Most of these effects have been found in the gametophytic system but plants with the sporophytic system also show end-season weakening. The sporophytic system, with its localized action on the stigmatic surface, shows bud self-fertility which indicates that the S stylar substance is not developed until the flower matures [85]. Removal or damage of the stigma mechanically [86], or by compatible pollen or species foreign pollen [87], or even by tryphine from compatible pollen of the same species or from pollen of another species [88], removes the barrier to selfing and can give almost the same seed set as with compatible pollination. All these environmental effects strengthen the basic dogma that the primary reaction is a positive oppositional one, and not a lack of a growth factor or stimulant.

4.4.5 Condition of the pollen and style during the reaction

Conditions of the style and pollen at the time of pollination and pollen tube growth affect the strength of the incompatibility reaction. Temperature in some gametophytic species and in heteromorphic species increases the strength of the reaction by a greater inhibition of selfed pollen tubes [89]. The Q_{10} in *Oenothera organensis* is 2.5, in *Prunus avium* 2.0, and in the two distylic *Primula* species 3.0 and 3.3 [28]. A similar effect has been found in *Petunia hybrida* [90]. Not all species show the temperature effect and even some genotypes of *Oenothera organensis* are completely temperature insensitive, the difference being due to an independent modifier gene **M**.

The temperature effect provides a means of showing that, at least in *Oenothera* and at 25 to 30°C, the incompatibility reaction is irreversible. Pollinations in the sensitive genotypes at 12°C give pollen tubes 120 mm long after 48 h; at 25 to 30°C the tubes stop growing at 2−4 mm after 30 min growth. When these 30°C pollinated styles are transferred to 12°C, there is no further growth [52].

A dose of 2000 rad or more of X-irradiation to the styles before or during the pollen tube growth in *Petunia* [91] and *Oenothera* (unpublished), seriously weakens the reaction and allows, in *Petunia,* selfed seed set and, in *Oenothera* at medium temperatures (c.20°C) only increased tube growth.

Neither the plant condition nor the condition of the pollen and style at the time of the reaction reveals anything of crucial importance for the elucidation of the reaction, but they offer explanations of pseudo-self-fertility which have been repeatedly reported from Darwin to the present day. Some of the conditions form the basis of techniques used in plant breeding to secure self-fertilization.

4.4.6 Preformation of the protein, localization and non-diffusibility in the style

An early experiment using style grafts in *Petunia* claimed to show an effect of the lower part of the style and ovary which was interpreted by assuming diffusion of incompatible specific substances from the ovary to the top of the style. Extensive graft experiments in *Oenothera* by Stirling Emerson [83] did not confirm this and showed that the reaction was highly localized. The *Petunia* results can now be explained by assuming that low molecular substances, such as sugars and amino acids, may be present in different amounts in grafted styles and hence influence the strength of incompatibility indirectly. That the S stylar protein is preformed and is not stimulated by the presence of the incompatible pollen was shown by reiterated pollination, first in *Petunia* [92] and later in *Oenothera* [71].

The presence of compatible and inhibited incompatible tubes side by side in the style, the grafting experiments and the reiterated pollinations, all point to a highly localized reaction.

CODA

Many models of incompatibility have been suggested from Darwin and Jost in the nineteenth century to the present day and, as very few have been refuted by facts, it is not intended to add yet another. Without being dogmatic and without making an exhaustive list, the accommodation of the following experimental facts appears to be absolutely essential for the homomorphic systems. They are essential because they have a restricted molecular interpretation and, with one exception, they are universal.

   (1) The multi allelic nature of the S gene.

   (2) The universal codominance in the style in the gametophytic system.

   (3) The dominance and negative complementation in diploid pollen in the gametophytic system.

(4) The I complementation between two genes in the two gene system.

(5) The protein–glucose–protein nature of the primary reaction.

(6) The fact of 3 and 4 gene systems, the frequency of which have yet to be shown.

SUMMARY

Plants do not have an immunological system comparable to anti-bodies and graft rejection of animals. Plants do not produce anti-bodies to invading pathogens or proteins; successful grafting between different genotypes of the same species and also between different species and even different genera is possible. They do possess, however, highly specific recognition systems between the male pollen and the female style. Their function is to inhibit the pollen of self or of sibs, with the same recognition genotype, before fertilization. They are outbreeding mechanisms of sophisticated efficiency not matched in any other group of organisms. These efficient systems are developed fully only in higher plants because there is a unique diploid somatic barrier, the style between the male and female gametes.

The genetical control of four main incompatibility systems, each with two or three subtypes, is well known. Each system has a characteristic pattern of dominance and codominance between alleles which is suited to other features of the system such as the genetic control of the pollen. An important feature of three of the systems is the large number of alleles. This implies that proteins may serve as the active principles. Biochemical and immunological evidence supports this but the exact nature of the inhibiting reaction is unknown. The reaction is highly localized, extremely specific and occurs at the surface of the membrane of pollen and stylar cells. Their widespread occurrence and the fact that families usually have only one system points to their great antiquity in the evolution of higher plants.

REFERENCES

1. East, E.M. (1931), *Harvey Lect.*, 1930–31.
2. Williams, W. (1947), *J. Genet.* **48**, 69–79.
3. Bateman, A.J. (1947), *Nature, Lond.*, **160**, 337.
4. Heslop-Harrison, J. (1971), *Pollen Development and Physiology*, Butterworth, London, pp. 75–98.
5. Vuilleumier, B.S. (1966), *Evolution*, **21**, 210–226.

6. Crowe, L.K. (1964), *Heredity,* **19**, 435–457.
7. Dulberger, R. (1975), *Proc. R. Soc. Lond. B.,* **188**, 257–274.
8. Ernst, A. (1936), *Z. indukt. Abstamm.-u. Vererblehre* **71**, 156–230.
9. Crosby, J.L. (1949), *Evolution,* **3**, 212–230.
10. Lewis, D. (1975), *Proc. R. Soc. Lond. B.,* **188**, 247–256.
11. Levin, D.A. (1968), *Am. Nat.* **102**, 427–441.
12. Dickinson, H.G. and Lewis, D. (1974), *Ann. Bot.* **38**, 23–29.
13. Dulberger, R. (1974), *Am. J. Bot.* **61**, 238–243.
14. Baker, H.G. (1966), *Evolution,* **20**, 349–368.
15. Knox, R.B. and Heslop-Harrison, J. (1971), *J. Cell Sci.* **9**, 239–251.
16. Heslop-Harrison, J. (1968), *New. Phytol.* **67**, 779–786.
17. Dickinson, H.G. and Lewis, D. (1973), *Proc. R. Soc. Lond. B,* **184**, 149–165.
18. Darwin, C. (1877), *The Different Forms of Flowers on Plants of the Same Species,* John Murray, London, p. 352.
19. Bateman, A.J. (1952), *Nature,* **170**, 496.
20. Brewbaker, J.L. (1959), *Indian J. Genet. Pl. Breed.* **19**, 121–133.
21. Pandey, K.K. (1960), *Evolution,* XIV, 98–115.
22. Mather, K. and de Winton, D. (1941), *Ann. Bot. New Series* V, 287–311.
23. Dowrick, V.P.J. (1956), *Heredity,* **10**, 219–236.
24. Dickinson, H.G. and Lawson, J. (1975), *Proc. R. Soc. Lond. B.,* **188**, 327–344.
25. Roggen, H.P.J.R. (1972), *Euphytica,* **21**, 1–10.
26. Ockenden, D.J. (1972), *New Phytol.* **71**, 519–522.
27. Dickinson, H.G. and Lewis, D. (1975), Biology of the Male Gamete. Ed. J.G. Duckett and P.A. Racey, *Linnean Soc.* 7, pp. 165–175.
28. Lewis, D. (1954), *Adv. Genet.* VI, 235–285.
29. Darlington, C.D. and Mather, K. (1949), *Elements of Genetics,* George Allen and Unwin, London.
30. Baker, H.G. (1961), *Rhodora,* **63**, 229–235.
31. Barlow, N. (1923), *J. Genet.* **XIII**, 132–146.
32. Mulcahy, D.L. (1964), *Am. J. Bot.* **51**, 1045–1050.
33. Ornduff, R. (1964), *Am. J. Bot.* **51**, 307–314.
34. Fyfe, V.C. (1956), *Nature,* **177**, 942–943.
35. Von Ubish, G. (1926), *Biol. Zbl.,* **46**, 633–645.
36. Esser, K. (1953), *Z. indukt. Abstamm.-u.Vererblehre,* **85**, 28–50.
37. Bateson, W., Saunders, E.R., Punnett, R.C. and Hurst, C.C. (1905), *Rept. Evol. Comm. R. Soc.* II, 154.
38. Demerec, M. (1924), *Am. J. Bot.* **11**, 461–464.
39. Brink, R.A. and Macgillvray, J.H. (1924), *Am. J. Bot.* **11**, 465–469.
40. Lewis, D. (1949), *Heredity,* **3**, 339–355.
41. Lundqvist, A. (1954), *Hereditas, Lund.* **40**, 278–294.
42. Heyman, D.L. (1956), *Aust. J. Biol. Sci.,* **9**, 221–231.
43. Lundqvist, A. (1975), *Proc. R. Soc. Lond. B.,* **188**, 235–245.
44. Lewis, D. and Modlibowska, I. (1942), *J. Genet.* **43**, 211–222.

45. Stout, A.B. and Chandler, C. (1941), *Science,* **94**, 118.
46. Lewis, D. (1947), *Heredity,* **1**, 85–108.
47. Lewis, D. (1951), *Heredity,* **5**, 399–414.
48. Pandey, K.K. (1956), *Genetics,* **41**, 327–343.
49. de Nettancourt, D. (1972), *Genetica Agraria* **XXVI**, 163–216.
50. Brewbaker, J.L. and Natarajan, A.T. (1960), *Genetics,* **45**, 699–704.
51. Pandey, K.K. (1967), *Heredity,* **22**, 255–284.
52. Lewis, D. (1960), *Proc. R. Soc. Lond. B.,* **151**, 468–477.
53. Wright, S. (1960), *Biometrics,* **16**, 61–85.
54. Fisher, R.A. (1961), *Aust. J. Biol. Sci.,* **14**, 76.
55. Denwood, T. (1963), *Hereditas, Lund.* **49**, 189–334.
56. de Nettancourt *et al.,* (1975), *Proc. R. Soc. Lond. B.,* **188**, 345–360.
57. Pandey, K.K. (1970), *Genetica,* **41**, 477–516.
58. Klein, J. (1975), *Biology of the Mouse Histoincompatibility-2 Complex,* Springer Verlag, Berlin.
59. Lewis, D. (1963), Genetics Today, *Proc. XI. Int. Congr. Genet.*
60. Correns, C. (1912), *Festshw. med-naturwiss. Ges.,* Münster **84**, 186–217.
61. Hughes, H.V. and Babcock, E.B. (1950), *Genetics,* **35**, 570–588.
62. Gerstel, D.U. (1950), *Genetics,* **35**, 482–506.
63. Bateman, A.J. (1954), *Heredity,* **8**, 305–332.
64. Crowe, L.K. (1954), *Heredity,* **8**, 1–11.
65. Thompson, K.S. (Quoted in Ockenden, D.J.), (1975), *Incompatibility Newsletter* **5**, 82–84.
66. Verma, S.C. *et al.,* (1976), *Proc. R. Soc. Lond. B.* (in press).
66a. Lewis, D. (1976), *Proc. R. Soc. Lond. B.* (in press).
67. Knight, R. and Rogers, H.H. (1955), *Heredity,* **9**, 69–77.
68. Cope, F.W. (1962), *Heredity,* **17**, 157–182.
69. Pandey, K.K. (1960), *Am. Nat.* **XCIV**, 379–381.
70. Bateman, A.J. (1947), *Heredity,* **1**, 303–336.
71. Lewis, D. (1952), *Proc. R. Soc. Lond. B.,* **140**, 127–135.
72. Mäkinen, Y.L.A. and Lewis, D. (1962), *Genet. Res. Camb.* **3**, 352–369.
73. Lewis, D., Burrage, S. and Walls, D. (1967), *J. Exp. Bot.,* **18**, 371–378.
74. Linskens, H.F. (1960), *Z. Bot.,* **48**, 126–135.
75. Nasrallah, M.E. and Wallace, D.H. (1967), *Heredity,* **22**, 519.
76. Linskens, H.F. (1953), *Naturwissenschaften,* **1**, 1–3.
77. Linskens, H.F. (1959), *Ber. D. Bot. Ges.,* **LXXII**, 84–92.
78. Lewis, D. (1943), *Ann. Bot. N.S.* **VII**, 115–122.
79. Linskens, H.F. (1955), *Z. Bot.,* **43**, 1–44.
80. Tupy, J. (1961), *Biol. Plantarum Prahe,* **3**, 1–14.
81. Linskens, H.F. (1975), *Proc. R. Soc. Lond. B.,* **188**, 299–311.
82. Linskens, H.F. and Esser, K. (1957), *Naturwissenschaften,* **44**, 1–2.
83. Emerson, S. (1940), *Bot. Gaz.* **101**, 890–911.
84. Tetsu Nakarishi and Kokichi Hinatu (1973), *Plant and Cell Physiol.* , **14**, 873–879.

85. Kakizaki, Y. (1930), *Jap. J. Bot.*, **5**, 133–208.
86. Tatebe, T. (1939), *J. Hort. Ass. Jap.*, **10**, 62–65.
87. Gerstel, D.U. (1950), *Genetics*, **35**, 666.
88. Roggen, H.P.J.R. (1975), *Incompatibility Newsletter*, **6**, 80–86.
89. Lewis, D. (1942), *Proc. R. Soc. Lond. B.*, **131**, 13–26.
90. Straub, J. (1958), *Z. Bot.* **46**, 98–111.
91. Linskens, H.F. (1960), *Naturwissenschaften* **47**, 547.
92. Straub, J. (1947), *Z. Naturforsch,* **2b**, 433–444.

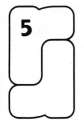

Catecholamine Receptors

ALEXANDER LEVITZKI
Department of Biological Chemistry,
The Hebrew University of Jerusalem,
Jerusalem, Israel

Receptors and Recognition, Series A, Volume 2
Edited by P. Cuatrecasas and M.F. Greaves
Published in 1976 by Chapman and Hall, 11 New Fetter Lane, London EC4P 4EE
© Chapman and Hall

5.1 INTRODUCTION

Almost every tissue undergoes a functional change in response to l-catecholamines ((−) catecholamines). In these tissues, a multitude of biochemical, physiological and pharmacological effects are induced by the catecholamines. The catecholamines involved are adrenaline (epinephrine), noradrenaline (norepinephrine) and dopamine. These small molecules act as either hormones originating from body fluids or as neurotransmitters released at an adrenergic synapse. In order to elicit a response, the catecholamines must first interact with specific receptors known as adrenergic receptors or adrenoreceptors. Therefore, a catecholamine capable of eliciting a response is known as an adrenergic agonist. This term applies not only to the naturally occurring catecholamines such as epinephrine and norepinephrine but also to synthetic catecholamines such as l-isoproterenol. An agent which blocks specifically the response elicited by an adrenergic agonist is known as an adrenergic antagonist or a blocker. Since adrenoreceptors have not been isolated and characterized, the characterization of these receptors relies mainly on the chemistry of the adrenergic ligands.

5.2 β- AND α-ADRENERGIC RECEPTORS

In 1906 Dale found [1] that preparations of the ergot alkaloid block the motor response of various organs to adrenaline and thus cause specific paralysis of the motor element. This same ergot preparation, however, did not affect the inhibitory response to adrenaline. Furthermore, inhibitory responses, not normally manifested by the organ, when exposed to adrenaline became apparent only after treatment with the ergot alkaloid. This early observation formulated the experimental basis for the existence of two major classes of adrenergic receptors. In another study, Dale investigated the increase in blood pressure and the response of some muscular organs to adrenaline and other phenylethylamine derivatives and found that these responses do not follow the same rules as those followed by the motor responses. He noted that the relative affinities towards the various phenylethylamines exhibited by the two

201

Table 5.1 Typical physiological actions of adrenergic receptors

System or tissue	Action	Receptor
Cardiovascular system, Heart	Increased force of contraction Increased rate	β
Blood vessels	Constriction Dilation	α β
Respiratory system, Trachel and bronchial smooth muscle	Relaxation	β
Iris (radial muscle)	Pupil dilated	α
Smooth muscle, uterus	Contraction	α
Spleen	Relaxation Contraction	β α
Bladder	Contraction Relaxation	α β
Skeletal muscle	Changes in twitch tension Increased release of acetylcholine Increased glycogenolysis	β α β
Adipose tissue	Increased lipolysis	β

A more detailed description of α and β receptor action is given in [5 and 6].

types of responses, differ, and that only the motor responses are blocked by ergot alkaloids. These two criteria formulated the basis upon which Ahlquist [2, 3] recognized the two main types of catecholamine receptors. In a detailed study, Ahlquist compared the relative efficacy of noradrenaline, adrenaline, isoprenaline and α-methyl derivatives of adrenaline and noradrenaline. He studied the effects of these catecholamine ligands on a wide variety of activities such as: vasoconstriction and vasodilation of different muscular beds, heart beat and the effects on the intestine and the uterus. Those effects were studied in cats, dogs and rabbits. Ahlquist found that there were two categories of response mediated by two distinct classes of receptors: α-receptors and β-receptors. The development of selective blockers for each type of receptor strengthened the conclusion that the adrenergic receptors fall into two main classes. In view of the complexity of the type of responses mediated by α and β-receptors (Table 5.1), it became apparent that the basis for their classification should be the chemistry of the stimulating agonist and of the blocking

Table 5.2 Ligand specificity of α-receptors and β-receptors

	(−)ISOPRENALINE		(−)EPINEPHRINE		(−)NOREPINEPHRWE		PHENYLEPHRINE
α	1	<	2	<	3	<	4
β	1	>	2	⪢	3	≫	4

antagonist. Thus the present characterization of adrenoreceptors [4, 5] is based on a two-fold procedure: (a) the relative potency (potency ratio) of a series of adrenergic agonists for eliciting the specific response and (b) the potency of an antagonist for blocking or inhibiting the response to a given agonist.

Based on these two principles Furchgott [4] gave a more general definition of α- and β-receptors:

β-receptors: a β-receptor is one which mediates a response pharmacologically characterized by: (1) a relative potency series: isoprenaline > adrenaline (epinephrine) > noradrenaline (norepinephrine) > phenylephrine and (2) a susceptibility to specific blockade by either propranolol or pronethalol at relatively low concentrations;

α-receptors: an α-receptor is one which mediates a response pharmacologically characterized by: (1) a relative potency series in which noradrenaline > adrenaline > phenylephrine ⪢ isoprenaline and (2) a susceptibility to specific blockade by phentolamine, dibenamine or phenoxybenzamine at low concentrations. A summary of the chemical formulas of α-antagonists and β-antagonists is given in Table 5.3. A summary of agonist specifcity of α and β-receptors is given in Table 5.2. Recently, receptors specific for dopamine have been identified, and as in the case of adrenaline receptors, two types of dopamine receptors

Table 5.3 α-blockers and β-blockers

seem to emerge (see below).

Primary biochemical signals coupled to adrenergic receptors: Of all catecholamine receptors the β-adrenergic receptors have been given the most study. This is probably due to the fact that the primary biochemical response induced upon agonist binding to the β-receptors has been identified and found to be the activation of the enzyme adenylate cyclase producing the second messenger cAMP [7, 8] from ATP:

$$ATP \xrightarrow{(-)\ catecholamine} cAMP + PPi \qquad (1)$$

The discovery made by Sutherland and his colleagues of cAMP and the formulation of the *second messenger* concept constitutes a milestone in the history of biochemistry in general and in hormone research in particular. The concept of cAMP being a second messenger arose originally from the studies of Sutherland and his colleagues on the effects of catecholamines [7, 8].

The β-receptor is believed to be coupled to the enzyme adenylate cyclase which uses intracellular ATP as the substrate. It seems that the relation between the β-receptor and adenylate cyclase is in principle similar to the relation between polypeptide hormones such as glucagon, ACTH, secretin etc. to the enzyme adenylate cyclase [10]. In a number of cell types, such as the fat cell and the liver cell, β-receptors as well as receptors to polypeptide hormones are found to be coupled to the enzyme adenylate cyclase. The 'second messenger' cAMP then triggers a cascade of biochemical reactions typical to the cell. Usually, the cascade begins with the activation of the enzyme protein kinase [11]. Claims have been put forward that β-adrenergic neurons act also via the formation of cAMP as a result of adenylate cyclase activation coupled to a β-type receptor at the post-synaptic membrane. Thus, for example, the inhibition of Purkinje cells by the noradrenergic nerve was found to involve increased levels of intracellular cAMP [12, 13]. More recently, claims have been put forward that the β-receptor is also coupled to biochemical signals independent of adenylate cyclase. It was shown in turkey erythrocytes [14] that $^{45}Ca^{2+}$ efflux is enhanced by β-agonists and is blocked by β-blockers. This cannot be mimicked by cAMP or dibutyryl cAMP. Recently, it was also shown [15] in turkey erythrocyte ghosts, that a specific GTPase activity is coupled to the β-receptor. This enzymatic activity may be involved in the regulatory control of the β-receptor dependent adenylate cyclase (see Section 5.4.1). Recently significant progress has been made in the elucidation of the biochemical responses coupled to α-receptors. It has been demonstrated that the primary event occurring upon occupation of the α-receptor by an α-agonist is the influx of Ca^{2+} which functions as the 'second messenger' [16]. Furthermore, the specific Ca^{2+} ionophore A 23187, when incorporated into the cell membrane, can substitute for the α-adrenergic ligand and bypass the receptor-dependent m mechanism [17]. The influx of Ca^{2+} as the primary event in the salivary gland (rat parotid) causes the efflux of K^+ ions with water [16, 18]. The efflux of potassium has also been recognized as an α-adrenergic effect in guinea pig liver [19, 20] and in adipose tissue [21].

In the case of dopamine receptors it was shown that in certain cases the enzyme adenylate cyclase is stimulated [22–24] and in other cases ion fluxes are probably involved [25]. These two distinct biochemical signals seem to reflect two types of dopamine receptors mentioned above. A summary of the primary biochemical signals elicited by the different catecholamine receptors is given in Table 5.4. A more detailed

Table 5.4 The primary biochemical events elicited by catecholamine receptors

Type of receptor	Biochemical event	References
β-Receptors	1. Activation of adenylate cyclase	7,8
	2. Ca^{2+}-efflux	14
	3. Activation of a specific GTPase	15
α-Receptors	1. Influx of Ca^{2+}	16–18
	2. Efflux of K^+	16–20
	3. Incorporation of phosphate to phosphatidylinesitol	99
Dopamine receptors	1. Activation of adenylate cyclase	22, 23
	2. Non-cyclase events*	

* H. Sheperd, personal communication

discussion on the coupling between the catecholamine receptor and the primary biochemical signal elicited, will be given in Section 5.4.1.

5.3 DIRECT PROBING OF CATECHOLAMINE RECEPTORS

One of the direct means of characterization of receptors is to perform binding studies, using specific ligands. Two general approaches can be used: (1) measurement of ligand binding using radioactive ligands or fluorescent ligands in order to measure both the quantity of receptors and their affinity towards the ligand; (2) use of irreversible affinity labels based on either agonist or antagonists. Until recently these direct approaches were unsuccessful in the catecholamine field. Only lately have techniques to study ligand binding to β-receptors reached an acceptable standard and satisfactory characterization of α-receptors or dopamine receptors using these approaches has not been reported as yet. Since most of the ligand-binding studies were performed on β-adrenergic receptors the major discussion will be devoted to this class of receptors. However, the principles emphasized in the discussion of ligand binding to β-receptors are equally valid for future studies on α-receptors or any other receptor.

Unlike the cholinergic system no rich source of adrenergic receptor is thus far available. As will be seen below, the quantity of β-receptors is in the picomoles per milligram range whereas in the electric organ of the electric fishes the amount of nicotinic receptor is in the nanomoles per milligram range. This state of affairs may explain, at least in part, the fact that the understanding of the nicotinic receptor at the molecular level is by far more advanced than the understanding of adrenergic receptors or any other known receptor.

5.3.1 The use of reversible agonists and antagonists

Until recently our understanding of the characteristics of the β-adrenergic receptor was based primarily on dose-response effects in whole animals, or isolated tissue preparations. The identification of adenylate cyclase and its capacity to be stimulated by β-adrenergic agonists, has allowed investigators to study the β-receptor under more controlled conditions, using the biochemistry of adenylate cyclase. Cell or membrane suspensions can now be studied by measuring the rate of cAMP formation subsequent to catecholamine stimulation. The following characteristics of the β-receptor have been formulated, based on the requirements for adenylate cyclase stimulation. The β-receptor exhibits ligand stereoselectivity towards both agonists and antagonists. In addition, the structure of the agonists and antagonists is similar and only the l-stereoisomers* are capable of activating or inhibiting the β-receptor–adenylate cyclase complex. The structural similarity between agonists and antagonists suggests that the two types of ligands compete for the same β-receptor site [26]. Indeed, the kinetics of propranolol inhibition of epinephrine-dependent adenylate cyclase activity in turkey erythrocytes established the fact that both ligands compete for the same site [27]. Based on structure–activity studies, it was found that the capacity of phenyl-ethylamine derivatives to stimulate the β-receptor adenylate cyclase strongly depends on the substitutents on the aromatic ring and on the aliphatic side chain. Although affinities for the different agonists and partial agonists, has not always been measured, their relative potency to stimulate adenylate cyclase is known [28]. It is generally believed that the receptor itself is a protein, although the possible importance of phospholipids in mediating the interaction between the receptor and the adenylate cyclase has been pointed out [29].

A serious limitation in studying the β-receptor by the measurement of adenylate cyclase activation, or any other metabolic process, is that the receptor–hormone interaction is not analysed directly. A direct measurement of catecholamine-β-adrenergic receptor interactions must involve studies of hormone binding.

* R in the notation of the organic chemists.

The binding of β-ligands to the β-receptor

Recently, the binding of ^{3}H-catecholamines to the β-adrenergic receptors in a variety of tissues has been reported [30–39]. A careful examination of those binding experiments revealed [27, 28, 40] that the ligand specificity found by these binding studies does not coincide with the ligand specificity as defined by their capacity to activate catecholamine-dependent adenylate cyclase in these systems. These discrepancies can be summarized as follows:

(1) Although (−) catecholamines stimulate adenylate cyclase [40], the previously reported studies reveal equal binding of the (−) as well as the (+) stereoisomers [31, 34, 38].

(2) A discrepancy between the concentration of propranolol required to inhibit catecholamine-stimulated adenylate cyclase and that required to inhibit catecholamine binding has been noted [20, 31, 34]. Only a small fraction of the bound catecholamines can be displaced by propranolol concentrations well above those fully inhibiting the epinephrine-dependent adenylate cyclase activity.

(3) The presence of a potent β-adrenergic agonist in the binding experiment, such as soterenol does not inhibit ^{3}H-catecholamine binding [40].

(4) The binding of ^{3}H-catecholamines is inhibited by catechol compounds which neither stimulate, nor inhibit, adenylate cyclase [31, 40].

(5) Fifty per cent displacement of bound ^{3}H-catecholamines has been found to occur only with a high concentration [31, 37] of added non-radioactive catecholamines, although theoretically, fifty per cent displacement should occur when the added non-radioactive ligand concentration equals the initial ^{3}H-ligand concentration beginning at a point of receptor saturation with the ^{3}H-ligand [28].

(6) Although activation of adenylate cyclase appears to be instantaneous, binding has been reported to require prolonged periods of incubation [31, 37].

From these considerations it appears that many of the conclusions based on these previous binding studies must be re-evaluated [40]. In fact, these studies probably measured binding of catecholamines to non-receptor catecholamine-binding proteins, such as catecholamine-O-methyl transferase, which are probably present in much larger amounts than the true β-adrenergic receptors. Thus, the reported number of receptors per cell, and the structural features required from a β-ligand are unrelated to the authentic β-adrenergic receptor. Similarly, the protein which was isolated and claimed to be the β-receptor [36], is not the true β-adrenergic receptor.

In an attempt to overcome these difficulties, we have taken another approach towards the study of ligand binding to the β-receptor [27, 28, 41]. We have taken advantage of the fact that propranolol binds primarily [27] and with high affinity to the β-receptor. We have studied ^{3}H-propranolol binding to turkey erythrocyte ghosts and its displacement by non-radioactive ligands. The following results were obtained [27, 28, 41]:

(1) Binding of propranolol and catecholamines is strictly specific for the ($-$)stereoisomers.

(2) The inhibition constant for propranolol as measured in the adenylate cyclase assay and the dissociation constant for propranolol binding are identical (2.5×10^{-9} M).

(3) All β-adrenergic antagonists displace ^{3}H-propranolol.

(4) Compounds which are neither agonists nor antagonists do not displace ^{3}H-propranolol.

(5) Fifty percent displacement of bound ^{3}H-propranolol by non-radioactive propranolol is defined only by the initial ^{3}H-propranolol concentration.

(6) All binding phenomena are rapid (less than 1 minute) and reversible.

This type of binding technique is the only method currently available which directly and exclusively measures binding to the β-receptor. It may now be applied to in-depth studies of the β-receptor-adenylate cyclase complex. This approach to the study of binding of radioactively labelled β-blockers to the β-receptors has since been adopted, independently, by other groups, who are also using radioactively labelled β-blockers for their studies. At the moment, three β-blockers are used for directly probing the β-receptor using binding assays: ^{3}H-propranolol [27, 28, 41, 44], ^{3}H-alprenolol [42] and ^{125}I-iodohydroxybenzylpindolol [43, 45–47]. All three blockers can only be displaced from the receptor by ($-$)catecholamines and not by (+) catecholamines. Similarly, compounds not affecting the β-receptor-dependent adenylate cyclase do not displace the antagonists from their binding site, although they compete effectively with catecholamine binding. Using ^{3}H-propranolol it was possible to show that the mature turkey erythrocyte possesses 600–1000 β-receptors per cell (about 20 receptors per μm^2 [27] as compared to 91 000 catecholamine binding sites which were wrongly assumed to be the β-receptors [31–38], as was pointed out earlier. Similar β-receptor density was found using ^{3}H-alprenolol [42, 48–50] and ^{125}I-iodohydroxybenzylpindolol [45–47] in different cell preparations. Recently we have shown that using ^{125}I-iodohydroxybenzylpindolol as a probe for the β-receptors in turkey erythrocyte membranes results in almost the same number of receptors per cell as that found originally using ^{3}H-propranolol [51].

5.3.2 Limitations of binding experiments

Using antagonists as probes for the β-receptors is not devoid of disadvantages. Thus, for example, the antagonists used are all hydrophobic in nature, in contrast to the catecholamine agonists. This hydrophobicity is responsible for a significant portion of the 'non-specific' binding of these ligands to the membranes. This 'non-specific' binding is caused largely by the solubility of these compounds in the lipid matrix of the membrane and is the major contributor to the linear, non-saturable component of the binding curve [28]. This component can be quantitated if the binding of the radioactive antagonist is measured in the presence of excess non-radioactive agonist or antagonist [27, 28]. The binding curve obtained in the presence of the excess non-radioactive ligand, must then be subtracted from the binding curve obtained in the absence of non-radioactive ligand, in order to obtain a quantitative measure of the specific receptors and their affinity towards the ligand.

5.3.3 Affinity labelling of the β-receptor

Affinity labelling of the β-receptor has recently been achieved by using a reversible β-blocker to which the ractive group bromoacetyl was attached [52–53]. The compound: N-(2-hydroxy-3-naphthoxypropyl)-N'-bromo-acetyl-ethylenediamine (Fig. 5.1) has been shown to inhibit irreversibly the epinephrine-dependent adenylate cyclase activity without damaging the F^- dependent activity in turkey erythrocyte membranes [52, 53]. Furthermore, propranolol and l-epinephrine offer protection against the affinity labelling reaction. Similarly the compound was shown to inhibit irreversibly the hormone-stimulated activity in a whole turkey red cell [53]. The loss of the epinephrine-dependent activity is accompanied by the loss of ^{3}H-propranolol binding [52, 53] demonstrating directly the loss of β-receptor subsequent to treatment with the affinity label. More recently, the ^{3}H-affinity label has been synthesized (Atlas and Levitzki, unpublished) and attempts are being made to characterize the protein moiety of the β-receptor. The availability of a β-receptor affinity label will also help to establish whether the hormone receptor and the enzyme activity reside on separate polypeptide chains. Since the β-receptor-activated enzymes from pigeon erythrocytes [54] and turkey erythrocytes [55] have already been solubilized by Lubrol PX and partially purified, an attempt has been made to use the ^{3}H-affinity label and identify the β-receptor in the solubilized state. This procedure can be of use since, in the solubilized form, the adenylate cyclase from pigeon erythrocytes

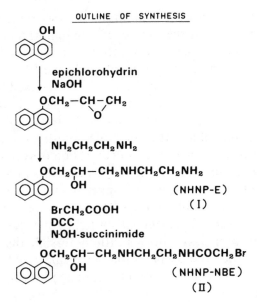

Fig. 5.1 An affinity label for the β-receptor. The outline of the synthesis of the compound is given.

or turkey erythrocytes is hormone-insensitive. The loss of hormone sensitivity of adenylate cyclase upon solubilization is well known, not only for catecholamine-stimulated adenylate cyclase, but also for adenylate cyclase activated by polypeptide hormones. This uncoupling event leaves the investigator with no direct means to monitor the hormone receptor unless an affinity label or a reversible ligand possessing high affinity are available.

5.3.4 Monitoring solubilized β-receptors

In principle reversible radiactive β-blockers can be used to monitor β-receptors in the solubilized state. Caron and Lefkowitz have reported [56] that ^{3}H-alprenolol was successfully used to monitor solubilized β-receptors. The authors [56] have used about 2×10^{-9} M receptor in their experiments where the dissociation constant towards the alprenolol is 1.0×10^{-8} M. It can be easily calculated that in an equilibrium dialysis experiment, the difference in radioactivity between the compartment possessing the membranes and the compartment containing the free ligand amounts to less than ten per cent. The solubility of the ^{3}H-alprenolol in the detergent, present in both compartments further

diminishes the sensitivity of the equilibrium dialysis experiment. It can
be easily demonstrated that the success of an equilibrium dialysis experiment depends on whether the concentration of receptors used is in the
range of the dissociation constant toward the ligand [28], conditions not
met in the study reported. Experiments we have designed to monitor the
β-receptor in the solubilized state using ^{3}H-propranolol have failed (Steer,
unpublished; Sevilla and Levitzki, unpublished). It was found that the
solubility of ^{3}H-propranolol in the butrol PX solution is one order of
magnitude higher than the maximal concentration of receptor accessible
experimentally. Thus an equilibrium dialysis experiment would result in
a small difference between two large numbers representing the concentration of free ^{3}H-propranolol plus bound ^{3}H-propranolol and the free
[3]H-propranolol, most of the signal contributed by the ^{3}H-propranolol not
bound to the β-receptor. It is however entirely feasible that compounds
such as ^{125}I-iodohydroxybenzylpindolol, possessing very high affinity to
the β-receptor, will be extremely useful for the monitoring of β-receptors
in the solubilized state.

5.3.5 Fluorescent probes for the β-receptors

Another approach to the study of the β-receptor uses a fluorescent
β-ligand which changes its fluorescent properties upon binding to the
β-receptor. Such a compound was recently synthesized (Fig. 5.2) and its

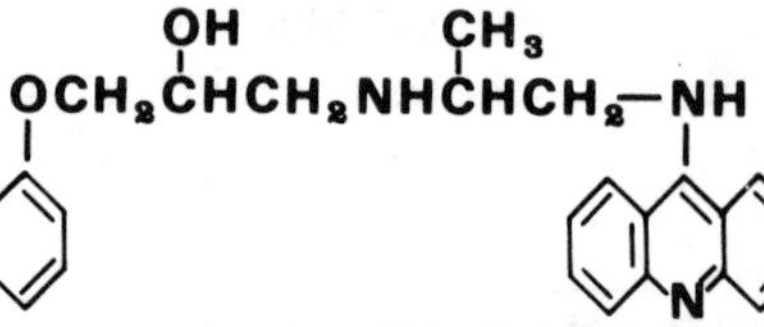

Fig. 5.2 A fluorescent β-blocker. The excitation peak is at 385 nm and the
emission peak at 420 nm. The compound was synthesized recently and was
shown to be a potent β-blocker [57]. Also, experiments with membrane
fragments of turkey erythrocytes enabled the β-receptor to be monitored
directly using this compound.

interaction with the β-receptor has been demonstrated in turkey
erythrocyte membranes [57]. This compound can be displaced from the
β-receptor only by l-epinephrine or l-propranolol and not by d-epinephrine
or d-propranolol. Using this compound, Atlas and her colleagues [58, 59]

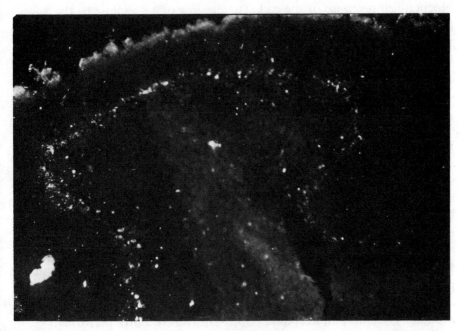

Fig. 5.3 Fluorescent micrograph of the sagittal section of the cerebellum.
An intense fluorescence is discerned at Purkinje cell layer subsequent to
the injection of the material to a mouse. This intense fluorescence is not
obtained if the mouse is first injected with l-propranolol but does appear
when the animal is first treated with d-propranolol [58]. With permission of
Nature and Drs. Melamed, Lahav and Atlas.

were able to map and localize β-receptors in various organs of the mouse
including the Purkinje cells (Fig. 5.3) in the cerebellum, the spinal cord,
the kidney, etc. This compound as well as similar fluorescent β-antagonists
and β-agonists can be of extreme importance in both the study of the
properties of the β-receptor per se and for the localization of β-receptor
in various tissues. Using this fluorescent β-blocker the localization of
β-receptors in heterogeneous tissues such as the brain, liver and kidney
can easily be achieved. This compound, therefore, offers a powerful tool
for the mapping of β-receptors.

5.4 ACTIVATION OF ADENYLATE CYCLASE BY β-AGONISTS

The occupancy of the β-receptor by a β-agonist is only the first step in
the production of the 'second messenger' cAMP which triggers the

biochemical response of the cell. The binding of a β-agonist to the β-receptor however, seems not to be the only event which controls the rate of cAMP production. Other regulatory ligands such as GTP [54, 60, 61] and Ca^{2+} [14, 62] were found to control the degree of enzyme activation.

5.4.1 The role of GTP and GppNHp in β-receptor activity

In the last few years it has been established, mainly through the studies of Rodbell and his colleagues that GTP and even more so GTP analogs such as $Gpp(CH_2)p$ and $Gpp(NH)p$ play a key role in the process of adenylate cyclase activation by glucagon. Rodbell *et al.,* suggested that GTP acts in a synergistic fashion with the hormone to activate the enzyme [63–65]. GTP analogs were found to play a similar role in the activation of adenylate cyclase by β-agonists in nucleated erythrocytes such as frog erythrocytes [61], turkey erythrocytes [60, 66, 67] and pigeon erythrocytes [54]. A detailed kinetic analysis of the synergistic activation of turkey erythrocyte adenylate cyclase by catecholamines and guanylylimidooliphosphate (GppNHp) was recently performed [66–68]. It was found that the role of the catecholamine hormone is to facilitate the activation of the enzyme by the guanyl nucleotide according to the following scheme:

$$RE \ + G \rightleftharpoons REG \tag{2}$$

$$REG + H \rightleftharpoons HREG \overset{k}{\to} HRE''G \tag{3}$$

where R is the receptor, E the enzyme, G the guanyl nucleotide effector, and H the hormone. The binding steps are fast and reversible and the rate of conversion of the inactive form of the enzyme to its active state E'' occurs with a rate constant of $k = 0.7$ min^{-1}. The active form of the enzyme system is extremely stable and remains permanently stable after removal of the hormone and the excess free GppNHp. This stable form of the enzyme can be converted back to the low activity form in the presence of hormone and ATP [66, 67, 69]. Neither hormone nor ATP alone cannot induce the deactivation of the enzyme and thus the reversal of the activation process requires the concerted action of the two ligands [69]. Both the activation process (Equation 3) and the deactivation process require a β-agonist and are blocked by a β-antagonist such as propranolol [66, 67]. Whether GppNHp effects on the membranes reflect the role of GTP is still not clear since GTP affects the system to a much lesser extent than GppNHp [70]. Some investigators invoke the argument that since GppNHp is resistent to phosphohydrolases, which are very active in this membrane preparation, its effect are pronounced in the cell-free membrane preparation. It still remains to be seen whether the GppNHp effect *in vitro* parallels the effect of GTP in the whole cell where the cyclase system and the various phosphohydrolase activities are presumably compartmentalized.

The properties of the β-receptor in terms of its affinity to β-blockers are not affected when the enzyme is in its active, E'' form. This has been demonstrated by direct binding measurements of ^3H-propranolol to the activated enzyme-receptor system as compared to the binding of ^3H-propranolol to the non-activated system [28]. Recently these results were confirmed using ^{125}I-iodohydroxybenzylpindolol [71].

5.4.2 The role of Ca^{2+}

Calcium is known to modulate the activity of hormone-dependent adenylate cyclase from different sources [9]. It was shown that Ca^{2+} inhibits the β-receptor dependent adenylate cyclase from turkey erythrocyte both in membrane preparations [62] as well as in whole turkey red cells [14]. In the latter case [14] Ca^{2+} inhibits the l-epinephrine dependent adenylate cyclase at 5 mM when the Ca^{2+} ionophore A23187 is incorporated into the red cell membrane. This procedure is necessary in order to overcome the very potent Ca^{2+} pump which operates in these cells. The nature of this Ca^{2+} inhibition was studied in detail on the membrane preparation [62] and it was found that the metal ion functions as a negative allosteric effector through a specific cooperative cluster of Ca^{2+} binding sites. The Ca^{2+} inhibitory sites, when occupied by Ca^{2+}, cause a complete inhibition of the adenylate cyclase activity. Thus Ca^{2+} functions as a pure V_{max} effector which has no effect on the affinity of the β-receptor towards agonists [62] or antagonists [28] and no effect on the kinetic parameters of the cyclase system towards Mg^{2+} or ATP [62]. Recently we have shown that a solubilized and partially purified adenylate cyclase retains its Ca^{2+} sensitivity [55]. The availability of such a preparation made it possible to study in further detail the parameters involved in the Ca^{2+} inhibition of adenylate cyclase. It was confirmed that the Ca^{2+} inhibition is brought about by the interaction of Ca^{2+} with specific Ca^{2+} regulatory sites on the adenylate cyclase moeity. Furthermore, Ca^{2+} is shown not to compete for the essential Mg^{2+} sites involved in the activation of adenylate cyclase.

As in the membrane preparation the soluble preparation, Mn^{2+} can support the cyclase reaction instead of Mg^{2+} [55]. The Mn^{2+} activated enzyme is however, insensitive to Ca^{2+} as was found for the membrane bound enzyme [62]. In cardiac tissue, on the other hand, Ca^{2+} was shown to inhibit adenylate cyclase by competition with Mg^{2+} at the Mg^{2+} allosteric site [72] Since Ca^{2+} function as the second messenger of α-receptor action, it may provide a link between α- and β-receptors in systems which possess both types of adrenergic receptors. Thus, since α- and β-receptors are antagonistic in their action whenever found in the same tissue (Table 5.1), it is not unreasonable that Ca^{2+} may assume this role. It is interesting that Batzri *et al.*, find that in the rat parotid gland, the α-blocker phentolamine slows down the fall in the level of cAMP subsequent to epinephrine stimulation, as compared to a system in the absence of the α-blocker [18]. This effect however may be due to secondary biochemical events other than the direct effect of Ca^{2+} on the level of adenylate cyclase activity. For example, Ca^{2+} is known to activate cAMP phosphodiesterase [73 and references therein] and thus an increase in intracellular Ca^{2+} may result not only in the inhibition of adenylate cyclase but also in the depletion of the cAMP pool. At this point, it can only be stated that the interaction of α-receptors with β-receptors is still not understood in biochemical terms and requires further investigation.

5.5 SELF REGULATION OF β-ADRENERGIC RECEPTORS

The availability of direct means for probing membrane receptors led to the discovery that the concentration of a ligand can regulate the concentration or the binding properties of the receptors on the target cell [74]. In the case of β-receptors it has been known for some time that β-adrenergic agonists can induce functional desensitization (tachyphylaxis, tolerance) of target tissues *in vivo* and *in vitro*. Using ^{3}H-alprenolol, Lefkowitz and his colleagues have shown that prolonged exposure of frog erythrocytes to β-adrenergic catecholamines *in vivo* [75] or *in vitro* [76] lead to a decrease of 50–70% in the number of alprenolol binding sites without a change in affinity towards the β-antagonist. The order of potency of the catecholamines is isoproterenol > epinephrine ≫ norepinephrine. These investigators also find that the β-blocker propranolol inhibits this action of agonists but does not by itself cause any decrease in the number of receptors. The down regulation of β-receptors differs from that of insulin or growth hormone in that the treatment *in vivo*

with the protein biosynthesis inhibitor cycloheximide does not prevent
recovery after down regulation [77]. These observations suggest that the
β-receptors are reversibly inactivated or masked and not lost as in the case
of insulin receptors and growth hormone receptors. Similar down regulation of β-receptors was recently reported by Axelrod and his colleagues.
The system studied by Axelrod *et al.,* are the β-receptors on the rat pineal
gland [49]. When these receptors were stimulated physiologically *in vivo*
by keeping the animals in the dark or pharmacologically by injecting
l-isoproterenol, a rapid fall in the number of ^{3}H-alprenolol binding sites
resulted. The fall amounts to 70% reduction within 2 hours subsequent
to stimulation. Within 4 hours a recovery in the number of ^{3}H-alprenolol
binding sites was found. Exposing the rats to light, thus decreasing the
sympathetic activity, results in the increase of the number of β-receptors
as measured by ^{3}H-alprenolol binding. These investigators also reported
[50] that the number of β-receptors on the pineal gland normally varies
with a circadian periodicity which is inversely related to the cycle of
neurotransmitter release [50]. Down regulation is an efficient mechanism
to regulate receptor response especially when the number of receptors is
such that only fractional occupancy of these receptors results in maximal
response. Under these circumstances, namely in the presence of 'spare
receptors', a decrease in the number of receptors does not cause a decrease in the potential *maximal response* but will shift the dose response
curve to higher agonist concentration. The reduction in the number of
β-receptors can in principle account for the functional desensitization of
target tissues to repeated agonist stimulation. However, it should be
stressed that other mechanisms such as the reduction in receptor affinity
may also be responsible for desensitization. Also, post-receptor biochemical mechanisms may operate in the phenomenon of desensitization.
Thus, for example, the number of opiate receptors in the brain is not
decreased by chronic exposure to high opiate levels [78, 79]. The effect
of the opiate ligand is the fast and reversible decrease in cAMP levels
where both the basal activity and the PGE-stimulated activity are inhibited. The cells compensate by a subsequent increase of the cAMP
levels, back to the level in the absence of the opiate ligand. In this case,
either the adenylate cyclase activity or its amount is increased [80]. The
removal of the opiate ligand results in an immediate further increase of
the cAMP levels, *above* the level typical of that in the absence of the
drug.

5.6 SPARE RECEPTOR CONTROL OF
CATECHOLAMINE-REGULATED PROCESSES

The level of circulating hormones such as catecholamines and glucagon
is well below the receptor hormone dissociation constant (Table 5.5).
The first biochemical signal induced by these two hormones is the biosynthesis of cAMP the 'second messenger', by the enzyme adenylate
cyclase. It is, therefore, clear that the activity of the enzyme is not fully
expressed. One can calculate (Table 5.5), that at the highest level of
hormone accessible physiologically [83–85] only 0.01% to 1.0% of the
total number of receptors become saturated. Although only a small
fraction of the receptors become saturated, it has been shown that the
biochemical processes elicited by the hormone are maximally stimulated

Table 5.5 Physiological hormone concentrations and their affinity towards receptors

Hormone	Physiological concentration (M)	Dissociation constant for receptor (M)
l-epinephrine and l-norepinephrine	1×10^{-9} to 3×10^{-8}*	6×10^{-6} to 1.0×10^{-5}†

* In chicken blood, epinephrine is 6 μg l^{-1} and that of norepinephrine is
   1.0 μg l^{-1} [81]; in human blood, epinephrine is 0.18–0.5 μg/ml^{-1} and
   norepinephrine is 2.4–4.4 μg/ml^{-1} [82]. Strenuous exercise or reduction in
   blood sugar results in up to a 10-fold increase in catecholamine concentrations
   [83–85].
† [27–28, 41].

under these conditions [86–88]. The existence of 'spare receptors' was
first implied in the response of many contractile tissue preparations to
histamine [89] or to muscarinic agonists such as acethylcholine [90–95].
In these cases, it has been claimed that receptors exist in a large excess
over the amount needed to bring about full response upon agonist binding. Until recently, however, it was not clear whether 'receptor reserve'
or 'spare receptors' really reflect a homogenous class of receptors in
excess, or a mixed population of binding sites, since rigorous criteria
were not used in the drug-binding measurements [27, 28, 40, 41]. Thus,
for example, it has been shown that catecholamines bind largely to non-
receptor binding sites [27, 40] and that β-adrenergic receptors constitute

only a small fraction of the total available catecholamine binding sites. Only when specific binding assays were developed [27–28, 41–43] and rigorous criteria [28, 40–41] for the definition of specific binding were used, did it become possible to measure quantitatively the number of β-adrenergic receptors in a number of tissues [27, 28, 42–50]. Therefore, in a selected number of cases, one can determine the exact meaning of 'spare receptors' in quantitative terms.

Let us consider the equation which governs the level of occupied receptor. The concentration of occupied receptor (HR) is given by:

$$[HR] = \frac{[H]_{free} \times [R_0 - [HR]]}{K_H} \tag{4}$$

where R_0 is the total concentration of receptors, namely:

$$\{HR\} = \frac{[R_0] \times [H]_{free}}{K_H + [H]_{free}} \tag{5}$$

Under conditions where the hormone concentration is well below its dissociation constant, namely, $K_H \gg [H]_{free}$, one obtains:

$$[HR] = \frac{[R_0] \times [H]_{free}}{K_H} \tag{6}$$

Equation 6 states that the concentration of occupied receptors is a linear function of the total concentration of receptors. Consequently, the extent of primary biochemical signal induced by the hormone, is directly proportional to the number of receptors at the target cell. Knowing the turnover number of the l-epinephrine-dependent adenylate cyclase on a turkey erythrocyte [28], one can calculate the maximal concentration of cAMP which can be attained once all the receptors are occupied. Since a turkey erythrocyte possesses about 1000 receptors per cell [27], and assuming that each receptor is coupled to a single adenylate cyclase catalytic unit [27, 28], one can calculate the amount of cAMP produced per minute (Table 5.6). It can immediately be seen that occupancy of less than 0.05% of the receptors will result in the production of cAMP concentration which exceeds significantly the dissociation constant for cAMP of the cAMP-dependent processes, within less than 1 min after exposure to to the hormone. Thus, for example, the known regulatory subunit of the cAMP dependent protein kinase binds cAMP with a dissociation constant of about 2×10^{-9} M to 2×10^{-8} M [96]. Therefore, the formation of 10^{-8} M to 10^{-7} M cAMP inside the cell, will result in the saturation of

Table 5.6 The maximal concentration of cAMP which can be produced in a turkey erythrocyte

Number of receptor cyclase molecules per cell	Number of cAMP molecules produced per cell per minute	Maximal concentration of cAMP per min attainable (M)‡	Concentration of cAMP per min produced at 0.016% receptor occupancy (M)
1000	1.2×10^5 *	8×10^{-7}	1.88×10^{-8}
1000	1.4×10^6 †	1.5×10^{-6}	1.44×10^{-7}

* Based on a turnover number of 120 min^{-1} [27] in the absence of GppNHp

† Based on a turnover number of 1400 min^{-1} in the presence of GppNHp [28].

‡ The volume of the turkey erythrocyte is 25 μm^3 or 2.4×10^{-10} cm^3/cell or 2.4×10^{-13} l. This number was reached by the following experiments: 1 ml of whole turkey blood was spun down at 50 000 x g for 15 min. The packed cells constituted 45% of the whole blood whereas the intercellular volume was 1.1%. (Levitzki and Porath Füredi, unpublished) as determined by measuring the trapping of ^{14}C-carboxymethyl cellulose. The value of 2.4×10^{-10} cm^3/cell for the cell volume can also be arrived at by measuring the cell dimensions under the microscope.

cAMP-dependent protein kinase, which in turn will saturate the biochemical processes mediated via cAMP dependent protein kinase.

If one examines a number of catecholamine-stimulated processes mediated by cAMP, such as: l-epinephrine-stimulated sodium transported and K$^+$ influx in turkey erythrocytes [86, 87] and enzyme secretion from the parotid gland [88–89], one indeed observes that the biochemical signal is saturated well below the saturation of the corresponding l-catecholamine-dependent adenylate cyclase with respect to the hormone (Table 5.7). In both cases it was shown that the effects induced by the hormone β-receptor interactions can be mimiced by exogenous cAMP [86] or dibutyryl cAMP [87].

Analysing the above examples Table 5.7 reveals that at circulating levels of epinephrine only a small fraction of the β-adrenergic receptors become occupied. The question then arises: what is the function of the excess receptor on the cell surface?.

The fraction of receptors saturated at a certain hormone level is governed by equation 6, the absolute extent of hormone-induced signal will depend on the total number of receptors. Thus, the role of 'spare receptors' is to provide the necessary total number of receptors so that

Table 5.7 The $S_{0.5}$ values for hormone in cAMP processes

The cAMP-mediated biochemical process measured	$S_{0.5}$ for the hormone mediated effect (M)	$S_{0.5}$ for the hormone in the adenylate cyclase reaction (M)
Epinephrine-stimulated Na^+ outflux in turkey erythrocytes	1×10^{-8} *	6×10^{-6} *
Epinephrine-stimulated α-amylase secretion by the rat parotid gland	2×10^{-7} ‡	1×10^{-5} ‡

* [86, 87].
† [27, 28, 41, 86, 87].
‡ [88, 89].

the fraction of receptors saturated will provide the level of 'second messenger' necessary to saturate the cAMP dependent biochemical processes. It seems, therefore, that the terminology of 'spare receptors' is inadequate since all the receptors are an integral part in generating the response.

It seems that many hormone-mediated processes and neurotransmitter-mediated responses operate by a mechanism involving the interaction of a ligand possessing low affinity to the receptor, with a large excess of the receptor. The end result of such a situation is similar to that obtained when a high affinity ligand interacts with a smaller excess of receptor, since the signal always depends on the product of the two. However, there is a possible advantage to a mechanism in which the agonist has low affinity to the receptor, and therefore, the extent of the signal elicited becomes dependent on the total number of receptors per cell. Such a mechanism allows for a discriminatory action of the same hormone on different tissues possessing the same receptor. Thus whether or not different tissues will respond to the same hormone concentration, depends on the total number of receptors per cell in each of the target tissues. Catecholamines, glucagon and histamine are responsible for a multitude of biochemical processes in different tissues, and in all three cases, there is evidence for the existence of 'spare receptors'. We would therefore like to suggest that for ligands acting on a variety of tissues, the 'spare receptor' mechanism is likely to operate.

Two approaches can be used to analyse the existence of 'spare receptors'.

One is a quantitative comparison between receptor occupancy and the dose response curve. This approach can be used when quantitative measurements of receptor occupancy can be performed. Another approach which was originally used by Nickersen [90] is to block irreversibly a fraction of the receptors and determine whether full response can still be obtained. Using this technique, Nickersen was able to show that the blocking of 99% of the histamine receptors by an irreversible blocker shifts the dose response curve with respect to the histamine agonist to higher concentrations but does not reduce the maximal response.

5.7 BIOCHEMICAL SIGNALS ELICITED BY α-ADRENERGIC RECEPTORS

It was mentioned in Section 5.2 that the more well-defined adrenergic receptors are of the β-type. In contrast to the situation in β-receptors, no characterization of α-receptors by direct binding assays is available. Also the structural correlation between α-agonists and α-blockers is not clear whereas β-blockers have many common structural features with β-agonists. Furthermore, the primary biochemical signal(s) are much less well characterized in the case of α-receptors than in the case of β-receptors. As to the last point some interesting progress has been made in the past few years. One of the best studied systems [16 and references therein] is the rat parotid gland which possesses three classes of receptors: β-adrenergic receptors, α-adrenergic receptors and a cholinergic muscarinic receptor. All three receptors reside on the same secretory cell. Activation of the β-adrenergic receptor results in the formation of cAMP which is responsible for enzyme secretion. Ca^{2+} serves as the second messenger in the response of the α-adrenergic receptor and the muscarinic receptor, both of which lead to K^+ release and water secretion. The primary event of α-receptor activation is however the entry of Ca^{2+}. Selinger *et al.*, were able to demonstrate that the ionophore for divalent ions A-23187 [97] in the presence of Ca^{2+} mimicks the α-adrenergic response causing massive K^+ release and water secretion from rat parotid slices [17]. This elegant experiment establishes the foundation for future work on the nature of α-adrenergic response. As indicated in Section 5.2, α-receptors are involved in a variety of activities and it remains to be shown that in all these activities Ca^{2+} functions as the 'second messenger' as in the case of the parotid gland. Whereas the activation of β-receptors lead to the activation of the enzyme adenylate cyclase, the activated α-receptor apprently fulfills the function of an ionophore introducing Ca^{2+} into the cell. The influx of Ca^{2+} then causes, by an unknown mechanism, the efflux of K^+ and water. The action of an α-adrenergic agonist is therefore dependent on the presence of extracellular Ca^{2+} as was indeed documented by Selinger and his colleagues. It was also pointed out by Selinger *et al.,* [17] that similar considerations may apply to a number of other hormones and neurotransmitters, the action of which is Ca^{2+} dependent [98]. This hypothesis can be tested by the use of the Ca^{2+} ionophore A23187 as in the case of the parotid gland. The combination of Ca^{2+} and the Ca^{2+} ionophore represent an experimental bypass of the α-adrenergic receptor much in the same way that the synthetic butyryl derivatives of cAMP offer an experimental bypass for β-receptor action.

The efflux of K^+ is not the only biochemical response elicited by the activation of α-receptors in the parotid gland. Oron *et al.,* [99] have shown that incorporation of inorganic ^{32}Pi into phosphatidyinositol, in slices of the parotid gland, is induced by α-adrenergic stimulation. This biochemical event is shown to be unrelated to the K^+ efflux and water secretion also induced by α-receptor activation. Interestingly enough, the divalent cation ionophore A 23187 which introduces Ca^{2+} into the cell, thus causing K^+ release [17], has no significant effect on the incorporation of ^{32}Pi into phosphatidylinositol. Conversely, the α-receptor induced phospholipid effect [99] is maximal in the absence of Ca^{2+} in the medium when there is no K^+ release from the cell. In summary, it can be concluded that α-receptor activation leads to two parallel and independent biochemical events in the rat parotid gland: (a) increase in membrane permeability towards extracellular Ca^{2+} which enters the cell and causes K^+ release; (b) the same interaction with the α-receptor results in the increased incorporation of ^{32}Pi into acidic phospholipids. This latter response is Ca^{2+}-independent and in fact, maximally stimulated in its absence.

It would be noted that the phospholipid effect was shown also to be induced by the activation of the muscarinic receptor in the same preparation of the parotid gland [99]. Phospholipid effects were shown for acetylcholine receptors as well as in response to other stimuli in other tissues [99] and references therin], in all these cases. The physiological response, however, in all of these cases, was dependent on the presence of extracellular Ca^{2+} as in the case of the α-receptor-induced response described by Selinger *et al.* These observation tend to strengthen Selinger's assertion that the phospholipid effect and the Ca^{2+} dependent K^+ release are two independent biochemical responses to α-receptor stimulation as in some other hormone or neurotransmitter stimuli.

5.8 INTERCONVERSION OF α- AND β-RECEPTORS ?

Reports in the literature have suggested that α-receptors and β-receptors
are two allosteric configuration of the same receptor. The experiments
on which this hypothesis is based were performed on frog heart where it
has been claimed that α-receptors prevail at low temperatures and trans-
formed into β-receptors at higher temperatures. It was claimed [100–
102] that stimulation of cardiac rate and contractibility by catecholamines
has the properties of a classical β-adrenergic response when experiments
are performed at warm temperatures (25–37°C) and of α-adrenergic
response when experiments are performed at cold temperatures (5–15°C).
Caron and Lefkowitz [103] examined this hypothesis by looking at the
adenylate cyclase activity at a wide range of temperatures. These investigators examined dog heart, rat heart, frog heart, and frog erythtocytes.
In all of these cases, it was found that the adenylate cyclase is stimulated
by adrenergic ligands typical of β-receptors at a wide range of temperatures.
Furthermore, the adrenergic inhibitors affecting cyclase at a wide range
of temperatures were always of the β-type. α-Blockers had no effect on
adenylate cyclase over a wide range of temperatures. As Caron and
Lefkowitz point out [103] the studies claiming the α to β interconversion
were performed on an intact tissue [100–102] whereas the adenylate
cyclase measurements were performed on membrane fragments [103].
Thus is still remains possible that the interconversion of α and β-receptors
requires the intact cellular structure. The integrity of the cellular structure
may preserve the biochemical mechanisms which may be responsible for
α-receptor to β-receptor interconversion. In conclusion, the possibility
of α-receptor to β-receptor interconversion still remains in view of the
pharmacological experiments [100–102] although it does no, at present,
find any support from direct biochemical experiments.

5.9 CATECHOLAMINES AS NEUROTRANSMITTERS: MECHANISTIC CONSIDERATIONS

Both β-receptors and α-receptors are involved in adrenergic innervation.
Table 5.1 summarizes the activities brought about by α-receptor or
β-receptor activation. Some of these activities are due to adrenergic
stimulation where the catecholamine functions as a neurotransmitter
released from presynaptic vesicles within the adrenergic synapse, acting
at a post-synaptic adrenoreceptor. Whereas there is much biochemical

data concerning the action of adrenergic agonists as hormones at α- and β-receptors, little is known about their mechanism of action as neurotransmitters. It seems, however, that β-receptor activity at β-adrenergic synapses involve the activation of adenylate cyclase and α-receptor activity involves changes in ion permeability where the first event is the influx of Ca^{2+}. For example, certain cells in the central nervous system such as the Purkinje cells of the cerebellum are innervated by adrenergic neurons. The neurotransmitter acting at the Purkinje cells is noradrenaline. It was shown that the noradrenaline transmitter is responsible for the sustained depression of spontaneous firing of the rat cerebellar Purkinje cells [105–107]. It was suggested that this noradrenaline action is mediated by cAMP [12, 13, 105]. Indeed phosphodiesterase inhibitors such as papaverine were found to depress the firing of Purkinje cells [12]. In conclusion, it seems that intracellular cAMP levels influence noradrenergic neurotransmission. This conclusion actually classifies these noradrenergic receptors as the β-type. It is therefore possible that other adrenergic synapses, not yet characterized may involve receptors of the β-type which function via the formation of cAMP. The presence of β-adrenoreceptors in cerebral tissue has been confirmed by electrophysiological techniques [106–107] and confirmed by direct measurements of ^{3}H-propranolol binding [44]. Another example where noradrenaline acts as a neurotransmitter is the rat pineal gland. In this case the noradrenaline neurotransmitter is released from the sympathetic nerves [108–109] and stimulates adenylate cyclase within the pineal gland [110]. Studies on this system are reported in [50] and references therein. This latter case is another example where the noradrenaline neurotransmitter acts on a post-synaptic membrane of the β-type.

As discussed in Section 5.7 it was found that the activation of α-receptors causes changes in the ion permeability of cell membranes of several tissues including intestinal smooth muscle, guinea pig liver and adipose tissue [19]. It is likely that the ion selectivity of the permeability increase varies between tissues. Thus, for example, in the inhibition in longitudinal muscle of the intestine, mainly K^+ ions are involved. It is possible that the K^+ ion effects are secondary to Ca^{2+} influx as found in the parotid gland by Selinger *et al.*, (Section 5.7). Indeed, it was suggested by Triggle [26] that α-adrenergic receptors may initiate contraction of certain smooth muscle, mostly arteries, by releasing Ca^{2+} from an intracellular store such as the sarco-plasmic reticulum. This mechanistic aspect was proposed mainly on the basis of the finding that the contractile response of such muscles falls only slowly when external Ca^{2+} is withdrawn.

REFERENCES

1. Dale, H.H. (1906), *J. Physiol., Lond.,* **34**, 163.
2. Ahlquist, R.P. (1948), *Am. J. Physiol.,* **153**, 586.
3. Ahlquist, R.P. (1967), *Ann. N.Y. Acad. Sci.,* **139**, 549.
4. Furchgott, R.F. (1972), in *Handbook of Experimental Pharmacology*, XXXIII, (H. Blaschko and E. Muscholl, editors), Springer-Verlag, Berlin, pp. 283–335.
5. Jenkinson, D.H. (1973), *Brit. Med. Bull.,* **29**, 142.
6. Blaschko, H. and Muscholl, E., Editors, (1972), *Handbook of Experimental Pharmacology*, XXXIII, Springer Verlag, Berlin.
7. Sutherland, E.W., Oye, I. and Butcher, R.W. (1965), *Recent Progress in Hormone Res.,* **21**, 623.
8. Robison, G.A., Butcher, R.W. and Sutherland, E.W. (1968), *Ann. Rev. Biochem.,* **37**, 149.
9. Perkins, J.P. (1973), *Adv. Cyc. Nucl. Res.,* **3**, 1, and references therein.
10. Cuatrecasas, P. (1974), *Ann. Rev. Biochem.,* **43**, 169.
11. Walsh, D.A., Perkins, J.D. and Krebs, E.G. (1968), *J. Biol. Chem.,* **243**, 3763.
12. Siggins, G.R., Hoffer, B.J. and Bloom, F.E. (1971), *Brain Res.,* **25**, 535.
13. Gähwiler, B.H. (1976), *Nature,* **259**, 483.
14. Steer, M.L. and Levitzki, A. (1975), *Arch. Biochem. Biophys.,* **167**, 371.
15. Cassel, D. and Selinger, Z. *Biochim. Biophys. Acta.* (in press).
16. Schramm, M. and Selinger, Z. (1975), *J. Cyc. Nucl. Res.,* **1**, 181.
17. Selinger, Z., Eimerl, S. and Schramm, M. (1974), *Proc. Nat. Acad. Sci.,* **71**, 128.
18. Batzri, S., Selinger, Z., Schramm, M. and Robinovitch, M.R. (1975), *J. Biol. Chem.,* **248**, 361.
19. Haylett, D.G. and Jenkinson, D.H. (1972), *J. Physiol., Lond.,* **225**, 721.
20. Haylett, D.G. and Jenkinson, D.H. (1972), *J. Physiol., Londs.,* **225**, 752.
21. Girardier, L., Seydoux, G. and Clausen, T. (1968), *J. Gen. Physiol., Lond.,* **52**, 925.
22. Kebabian, J.W., Petzold, G.L. and Greengard, P. (1972), *Proc. Nat. Acad. Sci.,* **63**, 2145.
23. Clement-Comer, Y.C., Kebabian, J.W., Petzold, G.L. and Greengard, P. (1974), *Proc. Nat. Acad. Sci.,* **71**, 1113.
24. Miller, R.J., Horn, A.S. and Iversen, L.L. (1974), *Mol. Pharm.,* **10**, 751.
25. Shepard, H. personal communication.
26. Triggle, D.J. (1972), *Ann. Rev. Pharmacol.,* **12**, 185.
27. Levitzki, A., Atlas, D. and Steer, M.L. (1974), *Proc. Nat. Acad. Sci.,* **71**, 2773.
28. Levitzki, A., Sevilla, N., Atlas, D. and Steer, M.L. (1975), *J. Molec. Biol.,* **97**,
29. Rubaclava, B. and Rodbell, M. (1973), *J. Biol. Chem.,* **248**, 3831.
30. Schramm, M., Feinstein, H., Naim, E., Lang, M. and Lasser, M. (1972), *Proc. Nat. Acad. Sci.,* **69**, 523.
31. Bilezikian, J.P. and Aurbach, G.D. (1973), *J. Biol. Chem.,* **248**, 5577.
32. Marinetti, G.V., Ray, T.K. and Tomasi, V. (1969), *Biochem. Biophys. Res. Comm.,* **36**, 185.

33. Tomasi, V., Koretz, S., Ray, T.K., Dunnick, J. and Marinetti, G.V. (1970), *Biochim. Biophys. Acta*, **211**, 31.

34. Dunnick, J.K. And Marinetti, G.V. (1971), *Biochim. Biophys. Acta*, **249**, 79.

35. Lefkowitz, R.J. and Haber, E. (1971), *Proc. Nat. Acad. Sci.*, **68**, 1773.

36. Lefkowitz, R.J., Haber, E. and O'Hara, D. (1972), *Proc. Nat. Acad. Sci.*, **69**, 2828.

37. Lefkowitz, R.J., O'Hara, D. and Warshaw, J. (1973), *Nature, New Biology*, **244**, 79.

38. Lefkowitz, R.J., Sharp, G. and Haber, E. (1973), *J. Biol. Chem.*, **248**, 342.

39. DePlazas, S.F. and De Robertis, E. (1972), *Biochim. Biophys. Acta*, **266**, 246–254.

40. Cuatrecasas, P., Tell, G.P.E., Sica, V., Parikh, I., and Chang, K.J. (1973), *Nature*, **247**, 92–97.

41. Atlas, D., Steer, M.L. and Levitzki, A. (1974), *Proc. Nat. Acad. Sci.*, **71**, 4246.

42. Lefkowitz, F.J., Mukherejee, C., Coverstone, M. and Carlon, M.G. (1974), *Biochem. Biophys. Res. Comm.*, **60**, 703–709.

43. Aurbach, G.D., Fedak, S.A., Woodward, C.J., Polmer, J.S., Hauser, D. and Troxler, F. (1974), *Science*, **186**, 1223.

44. Nahorski, S.R. (1976), *Nature*, **259**, 488.

45. Brown, E.M., Hauser, D., Toxler, F. and Aurbach, G.D. (1976), *J. Biol. Chem.*, **251**, 1232.

46. Brown, E.M., Rodbard, D., Fedak, S.A., Woodward, C.J. and Aurbach, G.D. (1976), *J. Biol. Chem.*, **251**, 1239.

47. Maguire, M.E., Wiklund, R.A., Anderson, H.J. and Gilman, A.G. (1976), *J. Biol. Chem.*, **251**, 1221.

48. Alexander, R.W., Davis, J.N. and Lefkowitz, R.J. (1975), *Nature*, **258**, 437.

49. Kebabian, J.W., Zatz, M., Romero, J.A. and Axelrod, J. (1975), *Proc. Nat. Acad. Sci.*, **72**, 3735.

50. Romero, J.A., Zatz, M., Kebabian, J.W. and Axelrod, J. (1975), *Nature*, **258**, 435.

51. Tolkovsky, A. and Levitzki, A., submitted to *Biochemistry*.

52. Atlas, D. and Levitzki, A. (1976), *Biochem. Biophys. Res. Commun.*, **69**, 397.

53. Atlas, D., Steer, M.L. and Levitzki, A. (1976), *Proc. Nat. Acad. Sci.* (in press).

54. Pfeuffer, T. and Helmreich, E.J.M. (1975), *J. Biol. Chem.*, **250**, 867.

55. Hansky, E. and Levitzki, A., submitted to *Biochemistry*.

56. Caron, M.G. and Lefkowitz, R.J. (1976), *Biochem. Biophys. Res. Commun.*, **68**, 315–322.

57. Atlas, D. and Levitzki, A., *Proc. Nat. Acad. Sci.* (in press).

58. Melamed, E., Lahav, M. and Atlas, D. (1976), *Nature*, **261**, 420.

59. Atlas, D. *et al.*, in preparation.

60. Bilezikian, J.P. and Aurbach, G.D. (1974), *J. Biol. Chem.*, **249**, 157.

61. Schramm, M. and Rodbell, M. (1975), *J. Biol. Chem.*, **250**, 2232.

62. Steer, M.L. and Levitzki, A. (1975), *J. Biol. Chem.*, **250**, 2080.
63. Salomon, Y., Lin, M.C., Londons, C., Rendell, M. and Rodbell, M. (1975), *J. Biol. Chem.*, **250**, 4239.
64. Lin, M.C., Salomon, Y., Rendell, M. and Rodbell, M. (1975), *J. Biol. Chem.*, **250**, 4246.
65. Rendell, M., Salomon, Y., Lin, M.C., Rodbell, M. and Berman, M. (1975), *J. Biol. Chem.*, **250**, 4253.
66. Levitzki, A., Sevilla, N. and Steer, M.L. (1976), *J. Supramolec. Struc.*, **4**, 405.
67. Sevilla, N., Steer, M.L. and Levitzki, A. (1976), *Biochemistry*, (in press).
68. Tolkovsky, A. and Levitzki, A. (manuscript in preparation).
69. Sevilla, N. and Levitzki, A., (manuscript in preparation).
70. Cassel, D. and Selinger, Z. (unpublished work).
71. Spiegel, A.M., Brown, E.M., Fedak, S.A., Woodward, C.J. and Aurbach, G.D. (1976), *J. Cyclic Nuc. Res.*, **2**, 47.
72. Drummond, G.E. and Duncan, L. (1970), *J. Biol. Chem.*, **245**, 976.
73. Appleman, A.M., Thompson, W.J. and Russel, T.R. (1972), *Adv. Cycl. Nuc. Res.* **3**, (P. Greengard and G.A. Robison, editors), Raven Press, New York, pp. 65–98.
74. Ruff, M. (1976), *Nature*, **259**, 265.
75. Mukherjee, C., Caron, M.G. and Lefkowitz, R.J. (1975), *Proc. Nat. Acad. Sci.*, **72**, 1945.
76. Mickey, J., Tate, R. and Lefkowitz, R.J. (1975), *J. Biol. Chem.*, **250**, 5727.
77. Lefkowitz, R.J. *Recent Progr. Hormone Research* (in press).
78. Pert, C.B., Pasternak, G. and Snyder, S.H. (1973), *Science*, **182**, 1359.
79. Klee, W.A. and Streaty, R.A. (1974), *Nature*, **248**, 61.
80. Sharma, S.K., Niernberg, M. and Klee, W.A. (1975), *Proc. Nat. Acad. Sci.*, **72**, 590.
81. Holzbauer, M. and Sharman, D.F. (1972), in *Handbook of Experimental Pharmacology XXXIII*, Springer Verlag, Berlin, pp. 110–185.
82. Vendsalu, A. (1960), *Acta Physiol. Scand.*, **49**, suppl. 173, 1.
83. Von-Euler, U.S. and Luft, R. (1952), *Metabolism*, **1**, 528.
84. Von-Euler, U.S. and Hellner, S. (1952), *Acta Physiol. Scand.*, **26**, 183.
85. Elmadjian, F., Lawson, E.T. and Neri, R. (1956), *Clin. Endocr.*, **16**, 22.
86. Gardner, J.D., Klaeveman, H.L., Bilezikian, J.P. and Aurbach, G.D. (1974), *J. Biol. Chem.*, **249**, 516.
87. Gardner, J.D., Klaeveman, H.L., Bilezikian, J.P. and Aurbach, G.D. (1973), *J. Biol. Chem.*, **248**, 5590.
88. Schramm, M. and Naim, E. (1970), *J. Biol. Chem.*, **245**, 3225.
89. Batzri, S., Selinger, Z., Schramm, M. and Robinovitch, R. (1973), *J. Biol. Chem.*, **248**, 361.
90. Nickersen, M. (1956), *Nature*, **178**, 697.
91. Stephenson, R.P. (1956), *Brit. J. Pharmacol.*, **11**, 379.
92. Ariens, E.J. (1964), *Mol. Pharmacol.*, **1**, Academic Press, New York.

93. Furchgott, R.F. (1966), in *Advances in Drug Research*, (Harper and Simmonds, A.B. editors), Academic Press, London.
94. Furchgott, R.F. and Bursztyn, P. (1967), *Ann. N.Y. Acad. Sci.*, **144**, 882.
95. Van Rossum, J.M. (1966), *Adv. Drug. Res.*, **3**, 189.
96. Rang, H.P. (1966), *Proc. Roy. Soc. B.*, **164**, 488.
97. Gilman, A.G. (1970), *Proc. Nat. Acad. Sci.*, **67**, 305.
98. Reed, P.W. and Lardy, H.A. (1972), *J. Biol. Chem.*, **247**, 6970.
99. Rubin, R.P. (1970), *Pharmacol. Rev.*, **22**, 389.
100. Oron, Y., Löwe, M. and Selinger, Z. (1975), *Mol. Pharmacol.*, **11**, 79.
101. Kunos, G., Yong, M.S. and Nickersen, M. (1973), *Nature*, **241**, 119.
102. Kunos, G. and Szentivanyi, M. (1968), *Nature*, **217**, 1077.
103. Buckley, G.A. and Jordan, C.C. (1970), *Br. J. Pharmacol.*, **38**, 394.
104. Caron, M.G. and Lefkowitz, R.J. (1974), *Nature*, **249**, 258.
105. Lake, N. and Jordan, L.N. (1974), *Science*, **183**, 663.
106. Siggins, G.R., Hoffer, B.J. and Bloom, F.E. (1969), *Science*, **165**, 1018.
107. Straughan, D.W., Roberts, M.H.T. and Socieszik, A. (1968), *Yugoslav. Physiol. Pharmacol. Acta*, **4**, 145.
108. Hoffer, B.J., Siggins, G.R. and Bloom, F.E. (1971), *Brain Res.*, **25**, 523.
109. Taylor, A.M. and Wilson, R.W. (1970), *Experientia*, **26**, 267.
110. Brownstein, M.J. and Axelrod, J. (1974), *Science*, **184**, 163.
111. Deguchi, T. and Axelrod, J. (1973), *Mol. Pharmacol.*, **9**, 612.